Band 21 · Juni 1985

6

isotopenpraxis

AKADEMIE-VERLAG

ISSN 0021-1915 Isotopenpraxis, Berlin 21 (1985) 6, 193—228

EVP 9,— M

An unsere Autoren

Die Zeitschrift veröffentlicht Beiträge zu folgenden Themen: Kontrolle und Analyse radioaktiver und stabiler Isotope, Anwendung radioaktiver und stabiler Isotope in allen Gebieten der Forschung und Technik, der Industrie, Landwirtschaft und Medizin, Herstellung radioaktiver und stabiler Isotope sowie von markierten Verbindungen, kernelektronische Geräte für die Isotopenanwendung (Dickenmesser, Dichtemesser u. ä.) Detektoren für isotopentechnische Arbeiten und den Strahlenschutz in Isotopenlaboratorien, Meßtechnik und Strahlenschutz, Dosimetrie in Isotopenlaboratorien, kerntechnische Geräte für die Isotopenanwendung (Beton, Bestrahlungsanlagen u. ä.), Strahlenchemie, Radiochemie, ökonomische und perspektivische Fragen, Fragen der Weiterbildung.

1. Aufnahme von Manuskripten

Zur Veröffentlichung angenommen werden nur druckreife, noch nicht anderweitig publizierte Beiträge in deutscher, englischer oder russischer Sprache zum Themenbereich der Zeitschrift und zwar

Originalarbeiten
Kurze Mitteilungen
Aus der Praxis
Zusammenfassende Berichte
Tagungsberichte
Informationen.

Sind der Beitrag oder Teile daraus bereits veröffentlicht oder ist eine anderweitige Veröffentlichung vorgesehen, so muß dies der Redaktion mit genauer Angabe der Zweitveröffentlichung mitgeteilt werden. Mit der Annahme des Manuskriptes behält sich der Verlag das Recht der Vervielfältigung, Verbreitung und Übersetzung innerhalb der gesetzlichen Schutzfrist vor. Auszugsweiser Nachdruck oder photomechanische u. a. Wiedergabe sind an eine schriftliche Genehmigung des Verlages gebunden. Manuskripte, Zeichnungen und Photos von veröffentlichten Arbeiten werden nur auf den besonderen und bis zum Zeitpunkt der Veröffentlichung mitgeteilten Wunsch des Autors zurückgesandt.
Bei der unverlangten Einreichung von Manuskripten wird empfohlen, eine Befürwortung eines der Herausgeber oder Redaktionsausschußmitglieder beizufügen.
Manuskripte sind einzureichen bei der Redaktion, 1086 Berlin, Leipziger Str. 3–4.

2. Form der Manuskripte

Die Autoren werden gebeten:
2 Manuskripte, eines davon in Original-Maschinenschrift, einseitig und mit 2zeiligem Abstand einzureichen; gut leserliche Photokopien sind zulässig; auf dem 1. Blatt (unter Weglassung der Bezeichnung)
den Titel der Arbeit
Vorname-Initialen und Namen sämtlicher Autoren
gegebenenfalls die Institution, in der sie tätig sind
eine kurze Zusammenfassung der Arbeit von höchstens 15 Zeilen
Keywords in alphabetischer Reihenfolge und
als Fußnote die Anschrift der Institution anzugeben;
wenn möglich, auf gesonderten Blättern Namen, Titel der Arbeit und Zusammenfassung in den beiden anderen obengenannten Sprachen 2fach beizufügen;
Literaturhinweise (fortlaufend numeriert), Fußnoten, Tabellen, Abbildungen (in 2–3facher Veröffentlichungsgröße), Abbildungsunterschriften (getrennt von den Abbildungen mit Zeichenerklärungen, Quellenangaben, falls erforderlich, und gegebenenfalls mit einem Vermerk für besondere Wünsche bezüglich der Abbildungsgröße), private und Dienstanschrift der Verfasser mit akademischen Titeln und ausgeschriebenen Vornamen auf besonderen Blättern ans Ende des Manuskriptes zu stellen;
Wünsche für den Kolumnentitel (Seitenüberschrift) bekanntzugeben;
Absätze durch Einzüge von mindestens 3 Anschlägen kenntlich zu machen;
die Maßeinheiten des SI-Systems („Systeme International") zu verwenden;
Korrekturen im Manuskript unmittelbar an der zu korrigierenden Stelle anzubringen;

Formeln deutlich lesbar und möglichst groß einzutragen;
Farbabbildungen und nachträgliche Änderungen oder Ergänzungen in Abbildungen zu vermeiden;

hervorzuhebende Textstellen zu unterstreichen, und zwar halbfetten Satz mit einer geraden schwarzen Linie, kursiven Satz (Schrägschrift) mit einer blauen Wellenlinie (alle Eigennamen). Sperrungen gestrichelt;
besondere Buchstaben durch farbige Unterstreichung zu kennzeichnen, und zwar Formelzeichen in

Griechisch-Kursiv (α, β, ...) rot,
Kursiv (A, B, ...) blau
halbfetter Kursiv ($\boldsymbol{A}$, $\boldsymbol{B}$, ...) 2 Linien blau (Vektoren),
halbfetter Grotesk (**A**, **B**, ...) 2 Linien violett (Tensoren);

bei Literaturangaben auf Vollständigkeit zu achten. In nachstehender Reihenfolge müssen angegeben sein:

a) bei *Zeitschriften*: Vorname-Initialen und Namen sämtlicher Autoren, Titel der Zeitschrift (nach „Periodica Chimica" oder möglichst wenig gekürzt), Bandnummer (unterstrichen), Jahreszahl (in Klammern) Heftnummer (nur bei Zeitschriften, die jedes Heft mit S.1 beginnen) sowie die Seitenzahl, auf der die zitierte Arbeit beginnt.

b) bei *Büchern*: Initialen der Vornamen und Name des Autors (Herausgeber werden durch den Zusatz „ed." gekennzeichnet), voll ausgeschriebener Titel des Buches in der Originalsprache, evtl. Bandnummer, voll ausgeschriebener Name des Verlages, Nummer der Auflage (außer der ersten), Erscheinungsort(e), Erscheinungsjahr (hier nicht in Klammern setzen!) und evtl. Kapitelnummer, Seitenzahl u. ä.

c) bei *Sammelbänden, Tagungsberichten*: Es soll zusätzlich der Titel des zitierten Beitrags angegeben werden. Die Seitenzahl muß unbedingt angegeben werden. Aus dem Zitat muß hervorgehen, ob es sich um einen Tagungssammelband oder um einen noch nicht veröffentlichten Vortrag handelt. Zwischen Tagungsort und Erscheinungsort des entsprechenden Tagungsberichts ist zu unterscheiden.

3. Korrekturen

Der Autor erhält zwei Umbruchabzüge, von denen einer möglichst schnell korrigiert an die Redaktion zurückgereicht werden soll. Auf Wunsch des Autors kann die Korrektur auch von der Redaktion gelesen werden.

Durch nachträgliche Manuskriptänderungen verursachte Korrekturkosten gehen, sofern sie 10% der Satzkosten übersteigen, zu Lasten des Autors.

Der Autor erhält 50 Sonderdrucke kostenlos. Weitere Sonderdrucke können bis zur Rückgabe der Korrekturabzüge auf Kosten des Autors bei der Redaktion bestellt werden.

Bezugsmöglichkeiten

Bestellungen sind zu richten

- **in der DDR** an den Postzeitungsvertrieb unter Angabe der Kundennummer des Bestellenden oder an den AKADEMIE-VERLAG BERLIN, DDR-1086 Berlin, Leipziger Straße 3–4, PF-Nr. 1233;
- **im sozialistischen Ausland** an eine Buchhandlung für fremdsprachige Literatur oder an den zuständigen Postzeitungsvertrieb;
- **in der BRD und Berlin(West)** an eine Buchhandlung oder an die Auslieferungsstelle KUNST UND WISSEN, Erich Bieber OHG, Wilhelmstraße 4–6, D-7000 Stuttgart 1;
- **in den übrigen westeuropäischen Ländern** an eine Buchhandlung oder an die Auslieferungsstelle KUNST UND WISSEN, Erich Bieber GmbH, Dufourstraße 51, CH-8008 Zürich;
- **im übrigen Ausland** an den Internationalen Buch- und Zeitschriftenhandel; den Buchexport Volkseigener Außenhandelsbetrieb der Deutschen Demokratischen Republik, DDR-7010 Leipzig, Postfach 160; oder an den AKADEMIE-VERLAG BERLIN, DDR-108 Berlin, Leipziger Straße 3–4.

Zeitschrift „Isotopenpraxis"

Herausgeber: Zentralinstitut für Isotopen- und Strahlenforschung der Akademie der Wissenschaften der DDR.
Verlag: Akademie-Verlag Berlin, DDR-1086 Berlin, Leipziger Straße 3–4, PF-Nr. 1233; Fernruf: 2236221 oder 2236229; Telex-Nr.: 114420;
Bank: Staatsbank der DDR, Berlin, Kto.-Nr.: 6836-26-20712.
Chefredakteur: Prof. Dr. H. Hübner, Leipzig.
Anschrift der Redaktion der Zeitschrift Isotopenpraxis: Prof. Hübner, DDR- 7010 Leipzig, Karl-Marx-Platz, Hauptpostamt, PSF 370; Fernruf: 281018.
Veröffentlicht unter der Lizenznummer 1527 des Presseamtes beim Vorsitzenden des Ministerrates der Deutschen Demokratischen Republik.
Gesamtherstellung: VEB Druckerei „Thomas Müntzer", DDR-5820 Bad Langensalza.
Erscheinungsweise: Die Zeitschrift „Isotopenpraxis" erscheint monatlich. Die 12 Hefte eines Jahrganges bilden einen Band. Bezugspreis je Band 240,— DM zuzüglich Versandspesen; Preis je Heft 20,— DM. Der gültige Jahresbezugspreis für die DDR ist der Postzeitungsliste zu entnehmen.
Bestellnummer dieses Heftes: 1079/21/6.

AN (EDV) 12831
00900

Isotopenpraxis 21 (1985) 6, S. 193 bis 198

Deuterium and Oxygen 18 in Water Vapour and Precipitation: Application to Atmospheric Water Vapour Transport and to Paleoclimate*)

C. Sonntag, H. Schoch-Fischer

(Institut für Umweltphysik der Universität Heidelberg)[1])

Spatial and temporal variations of δD and δ¹⁸O in mean monthly precipitation and groundwater are described by means of a simple water vapour advection model, which links the isotope data to the vertically integrated horizontal flow of vapour. This model yields a temperature dependence of the isotope precipitation data, which is compared with the experimental temperature effect. Short-term variations of δD and δ¹⁸O in mean daily water vapour reflect the varying influence of different air masses on the local weather. A spectral analysis of these short-term variations yields predominant periodicities of about 10, 20 and 30 days, which are related to the long-wave pattern of the zonal drift in the upper troposphere. δD and δ¹⁸O in plant material (tree rings, peat), solids, lake sediments, speleothems and in other carbonates of paleoclimatic interest is primarily influenced by the isotopic composition of local precipitation, atmospheric moisture and by temperature dependent isotope fractionation. The relevance of such fossil material (peleowaters included) fo paleoclimate and in particular for the reconstruction of the water vapour circulation pattern in the past is discussed.

Räumliche und zeitliche Variationen von δD und δ¹⁸O in den Monatsmittelwerten der Niederschläge und im Grundwasser werden mit Hilfe eines einfachen Wasserdampf-Advektionsmodells beschrieben, das die Isotopendaten mit dem vertikalen Wasserdampfstrom verbindet. Das Modell liefert eine Temperaturabhängigkeit der Isotopendaten in den Niederschlägen, die mit dem experimentell ermittelten Temperatureffekt verglichen wird. Kurzzeitige Variationen der Tagesmittelwerte von δD und δ¹⁸O werden im Luftwasserdampf durch den Einfluß unterschiedlicher Luftmassen verursacht. Eine spektrale Analyse dieser kurzzeitigen Variationen ergibt signifikante Periodizitäten von 10, 20 und 30 Tagen, welche mit den langwelligen zonalen Drifterscheinungen in der oberen Troposhäre verknüpft sind. Die δD- und δ¹⁸O-Werte von Pflanzenmaterial, Bodenproben, Seesedimenten, Höhlenkalken und anderen Karbonaten werden durch die jeweiligen isotopen Zusammensetzungen der örtlichen Niederschläge, der Luftfeuchte und der temperaturabhängigen Isotopenfraktionierungen beeinflußt. Die Bedeutung dieser fossilen Materialien (einschließlich der Paläowässer) wird im Zusammenhang mit Aussagen zum Paläoklima und der einst herrschenden Luftzirkulation diskutiert.

Keywords

atmospheric precipitations; carbonates; climates; deuterium; groundwater; isotope ratio; meteorology; oxygen 18; stable isotopes; water vapour

1. Introduction

The IAEA/WMO Global Precipitation Network provides deuterium and oxygen 18 data of mean monthly precipitation over the past twenty years [1]. These concentrations of the heavy water molecule species HDO and $H_2{}^{18}O$ in ordinary water $H_2{}^{16}O$ are usually presented as permille deviations δD and $\delta^{18}O$ of the sample's isotope ratio $R = (HDO)/(H_2O)$ or $(H_2{}^{18}O)/(H_2O)$ from the isotope ratio of Standard Mean Ocean Water:

$$\delta D \text{ (or } \delta^{18}O) = (R/R_{SMOW} - 1) \times 10^3 \quad [‰] \tag{1}$$

The isotope data of precipitation shows space and time (mainly seasonal) variations, which are caused by isotope fractionation in the evapoaration and condensation processes along the atmospheric branch of the water cycle [2]. In the phase transitions involved the heavy water molecules HDO and $H_2{}^{18}O$ preferentially remain in or are transferred into the liquid phase. This behavior is expressed by the isotope fractionation factor

$$\alpha = R_{liq}/R_{vap} > 1 \tag{2}$$

The δD and $\delta^{18}O$ precipitation data are ranging from about +20‰ to −400‰ and from about +3‰ to −50‰ respectively. These predominantly negative δ-values are mainly due to progressive isotopic depletion of moist air masses, which are precipitating in the course of isobaric or moist adiabatic cooling along their trajectories from warm oceanic regions towards locations of lower temperatures. The isotope fractionation during the cloud droplet formation occurs at thermodynamic equilibrium between the vapour and liquid phase. The corresponding equilibrium fractionation factor α_e increases considerably with decreasing temperature, but the ratio between the fractionation $\alpha_e - 1$ of D and ¹⁸O varies from 7·7 at +40 °C to 9·6 at −20 °C only. This is the main reason why the isotope data of precipitation are fairly close to the "Meteoric Water Line" $\delta D = 8 \times \delta^{18}O + 10$ in the commonly used δD versus $\delta^{18}O$ presentation.

As being expected from the foregoing, the isotopic composition of precipitation depends primarily on the

[1]) Address: D-69 Heidelberg, Im Neuenheimer Feld 366

*) Presented at the 3rd Workshop on „Isotopes in Nature", November, 15–18, 1983 in Leipzig, GDR.

local temperature. This has been shown by a multiple linear regression analysis between the mean monthly isotope data of the IAEA/WMO Precipitation Network and the geographical position (latitude, altitude) of the stations, the temperatures, and the precipitation amounts there [3]. This temperature dependence, however, is not linear. At high temperatures (30 – 20 °C, tropical rains, mid-latitude summer rains) the temperature effect is $\Delta\delta^{18}O/\Delta T \cong 0{\cdot}2-0{\cdot}3^0/_{00}/°C$, but increases to $0{\cdot}5^0/_{00}/°C$ at 20 to 10 °C and finally to more than $1^0/_{00}/°C$ at temperatures below 0 °C (mid-latitude snowfall, high latitude precipitation). δD shows a similar, but 8 times higher absolute temperature effect.

In the early days of the IAEA/WMO Network the problem has already been discussed, in how far the isotope precipitation data can be used as tracer for the atmospheric water vapour circulation [4]. Recently this problem has drawn new attention, since δD and $\delta^{18}O$ of various materials are considered to be powerful indicators of paleoclimate. In this context the question arises, to which extent time variations of δD and $\delta^{18}O$ of dated paleowaters, plant materials (tree rings), speleothems etc. can be transferred into variations of the hydrometeorological parameters in the past (paleo-temperature, -humidity, -precipitation, -evaporation). The solution of this problem needs detailed information on the variation of δD and $\delta^{18}O$ in atmospheric water vapour in comparison with the variation of the daily weather parameters. Isotope precipitation data provide only an episodic picture of the rainy weather conditions, which may turn out to be insufficient for reliable conclusions on changes in the atmospheric water vapour circulation.

2. Spatial Variation of δD and δ¹⁸O-Water Vapour Advection Model

As having been mentioned above, precipitating air masses undergo a progressive isotopic depletion along their trajectories from high to low temperature locations. The δD-isoline presentation of modern European groundwater (Fig. 1) shows a west to east decrease of δD, which means, that Western and Central Europe receive rain from the Western drift. The higher isoline-densities at the western side of the Alps and the Scandinavian mountains are due to the altitude effect which superimposes the continental

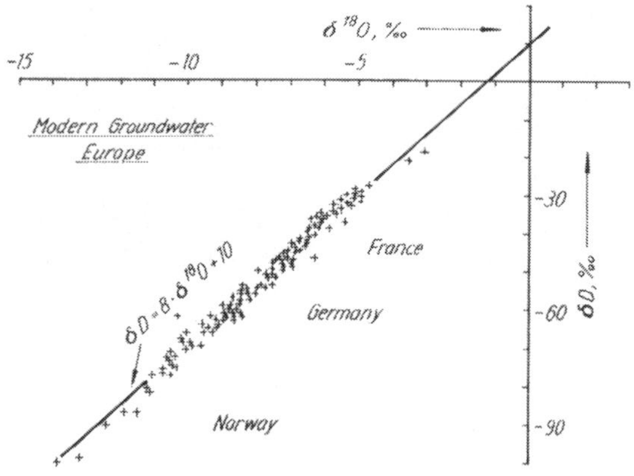

Fig. 1b. δD versus $\delta^{18}O$ diagram for modern European groundwaters The spread around the regression line is predominantly due to our present analytical precision of $\pm 1^0/_{00}$ for deuterium.

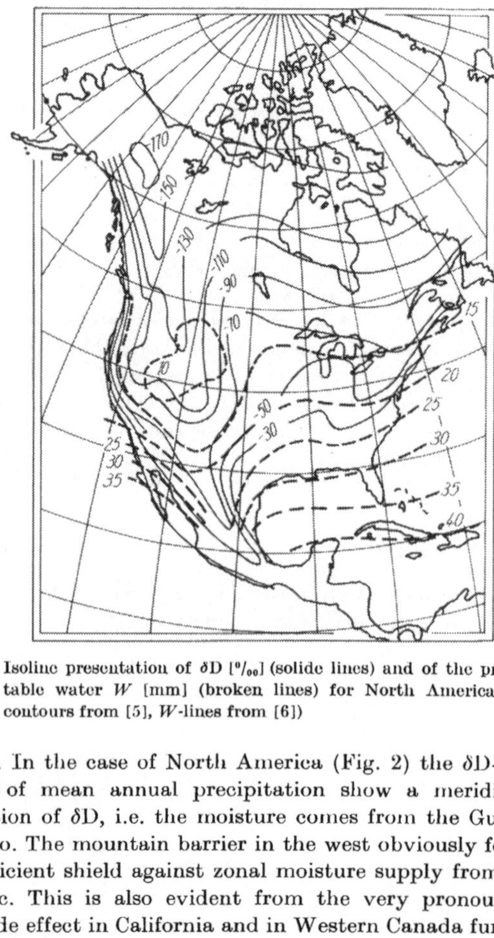

Fig. 2. Isoline presentation of δD [$^0/_{00}$] (solide lines) and of the precipitable water W [mm] (broken lines) for North America (δD-contours from [5], W-lines from [6])

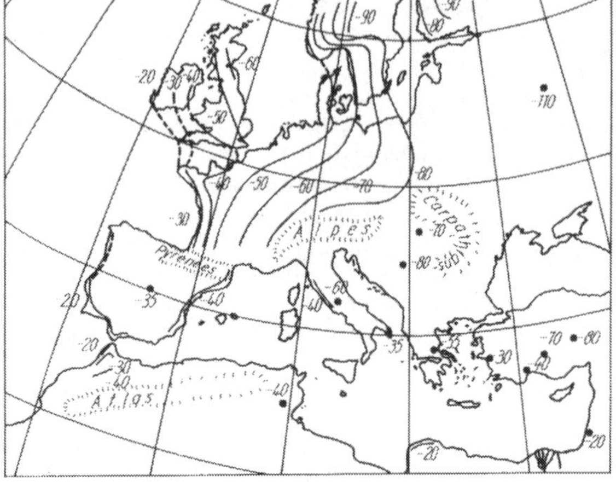

Fig. 1a. Spatial distribution of δD [$^0/_{00}$] in modern European groundwaters The data points (about 400) uniquely fall into distinct isozones, the resolution across the isolines (excluding the influence of altitude effects) may turn out to be not much more than 50 km or about $\pm 2^0/_{00}$ in δD.

effect. In the case of North America (Fig. 2) the δD-contours of mean annual precipitation show a meridional variation of δD, i.e. the moisture comes from the Gulf of Mexico. The mountain barrier in the west obviously forms an efficient shield against zonal moisture supply from the Pacific. This is also evident from the very pronounced altitude effect in California and in Western Canada further north. Fig. 2 also presents the contours of the precipitable water W, which pattern is quite similar to the one of δD.

The isotope data and the precipitable water can be linked by a fairly simple water vapour circulation model which considers the vertically integrated vapour flow across the continents [7]. Under the assumption of exponentially decreasing vertical profiles of the tropospheric moisture (water vapour mixing ratio) and its isotope ratio R, the trophospheric bulk moisture content W and its isotope ratio $R_w = R$ depend only on the ground level data, i.e. W is determined by the local temperature and relative humidity, R by the isotope data of local air moisture or precipitation. Under the further assumption of advective flow only with constant advection velocity (mean wind velocity at 850 mb), this model yields the *Rayleigh* formula

$$dR/R = (\alpha_e - 1) \times dW/W$$

or

$$d\delta/(1 + \delta) = (\alpha_e - 1) \times dW/W \quad (3)$$

which describes the isotopic depletion at decreasing atmospheric moisture. This simple advection model simulates the isotope pattern of North America and of Europe (W-contours not shown in Fig. 1) fairly well. It simulates also the seasonal variations of δD and δ^{18}O in the mean monthly precipitation at European network stations [8]. Moreover the *Rayleigh* formula above yields a fairly simple expression for the temperature dependence of the isotope data. The water vapour saturation pressure and thus the atmospheric moisture content and its vertical integral, the precipitable water W, vary approximately exponentially with the temperature T, $W(T) \sim \exp(T/T_0)$, where $T_0 \cong 15$ °C for the temperature range from 0 °C to 30 °C.

Insertion into eq. 3 yields

$$\Delta\delta/\Delta T = (\alpha_e - 1)/T_0 \quad (4)$$

With $\alpha_e(D) = 1.090$ a temperature dependence of $\Delta\delta\, D/\Delta T = 6.0‰/°C$ (0.75 for δ^{18}O) is obtained, which number compares fairly well with the one of *Dansgaard* [9] for the temperature effect in mean annual precipitation. However, the temperature variation of the mean monthly isotope data from the IAEA/WMO Precipitation Network (see previous section) and of the isotope records of mean daily water vapour (next section) is by a factor of 2 smaller within the temperature range considered. These smaller numbers support the idea of turbulent water vapour transport instead of the assumed advective flow. In the case of turbulent flow the temperature dependence it expected to be proportional to $\sqrt{\alpha_e} - 1$ instead of $\alpha_e - 1$ [4]. This discrepancy has not yet been solved. Finally it should be mentioned here, that European paleowaters of ^{14}C-groundwater ages > 10,000 years B.P. are isotopically depleted by about $\Delta\delta$ D = −10 to −15‰ and $\Delta\delta^{18}$O = −1.2 to −1.9‰ in comparison to the local modern groundwater

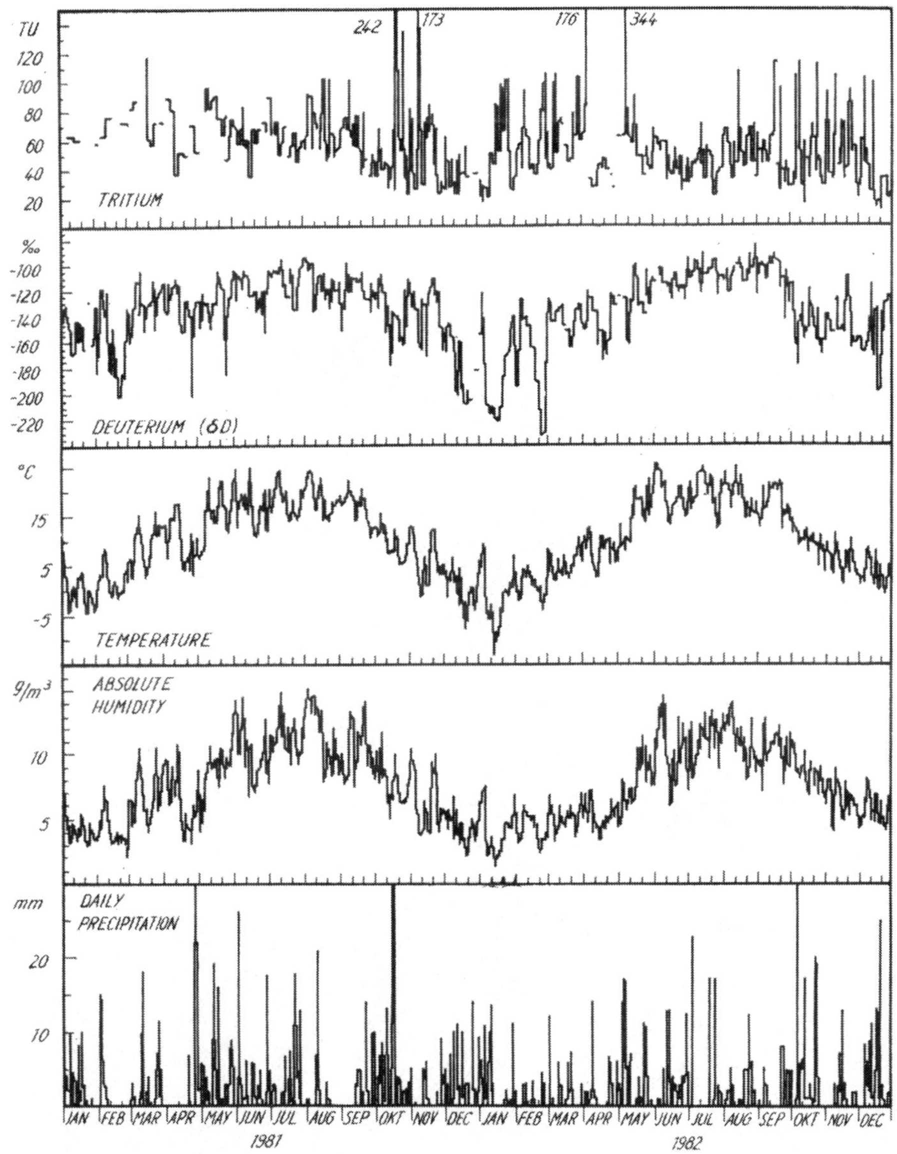

Fig. 3. Two years record of δD in mean daily atmospheric water vapour at Heidelberg together with time series of the local temperature, absolute humidity and precipitation as well, tritium time series not being discussed

[10]. If this depletion is attributed to the temperature effect, apleotemperatures can be derived for the last ice-age, which are 3 to 5 °C lower than the present local temperatures. This temperature data compares fairly well with the ones obtained from the noble gas contents (Ar, Kr, Xe) of these waters [10].

3. δD and $\delta^{18}O$ in Mean Daily Atmospheric Vapour

Further improvement of our present knowledge of the hydrometeorological and climatic information contained in the isotope data of precipitation needs continuous records of the short-term variations of δD and $\delta^{18}O$ in atmospheric water vapour, which have to be compared with the variations in the local weather data. Fig. 3 shows a two-years record of δD in mean daily water vapour at Heidelberg together with the local temperature and absolute humidity, which represents the precipitable water, and with the local precipitation [11]. The strong short-term variations of these time series, which are superimposed on the seasonal variation, reflect the varying contribution of tropical and polar, marine and continental air masses to the mid-latitude weather conditions. In several cases extremely low δ-values are related to high precipitation rates (amount effect). We have tried to simulate the daily isotope data by reconstructing the "(isotope) rainout-history" of individual air masses along their trajectories by means of the horizontal vapour advection model. Although the agreement between these model predictions and the observed isotope data were very poor in many cases, the results have encouraged us to begin a synoptic water vapour investigation with sampling stations at Miami (representing the source region of European air moisture), Paris, Hannover, Heidelberg and Cracow. Fig. 4 shows δD-records of these stations for the first months of 1983 together with typical weather situations, which qualitatively explain the isotope data at these particular days [11]. We are, of course, aware that this investigation does not cover the time variations of δD and $\delta^{18}O$ in the vertical dimension, a lack of information which may turn out to be a serious limitation in this isotope study.

Going back to Fig. 3, we find that spectral analysis of this two year's isotope record of air moisture at Heidelberg yields a red noise spectrum with prominent periodicities of about 10, 20 and 30 days, with high statistical significance. Global weather chart inspection shows that these periodicities are related to the long wave pattern of the zonal drift in the upper troposphere (500 mb). These Rossby Waves, which are known to steer the cyclogenesis in the lower troposphere (moist layer), move in the zonal

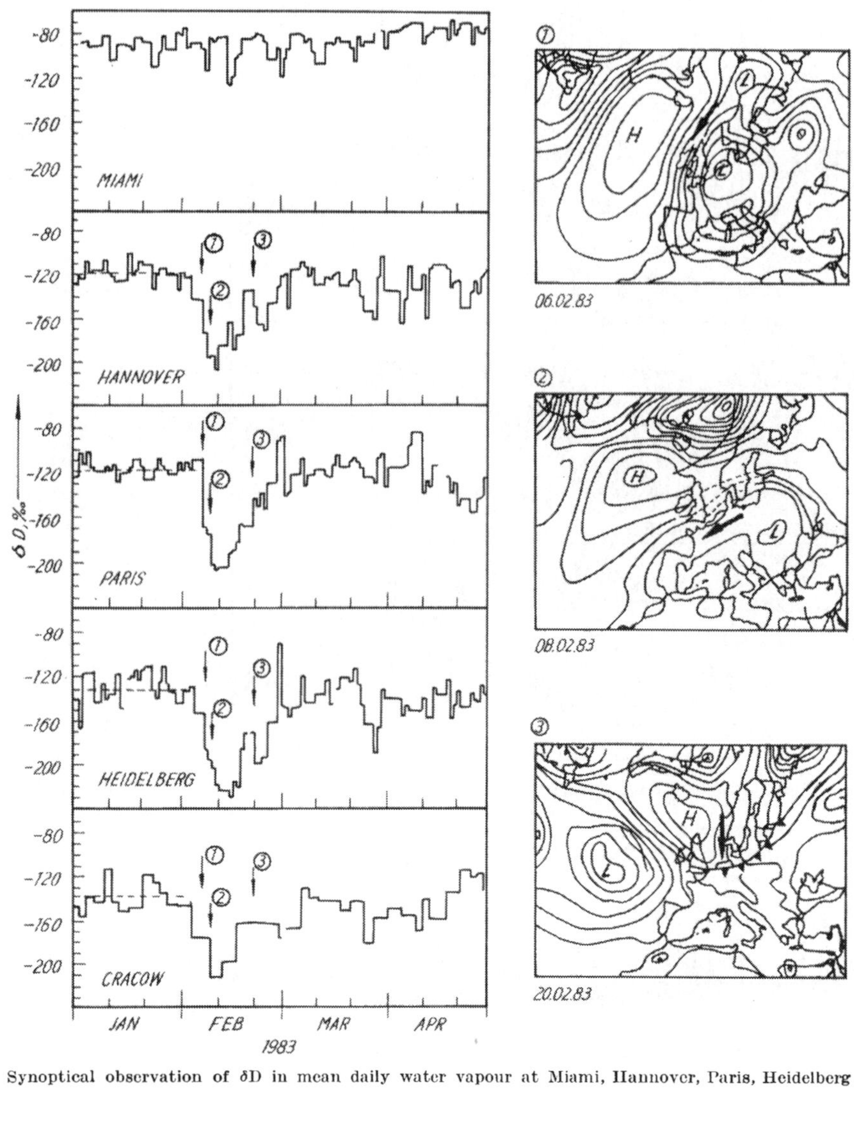

Fig. 4. Synoptical observation of δD in mean daily water vapour at Miami, Hannover, Paris, Heidelberg and Cracow

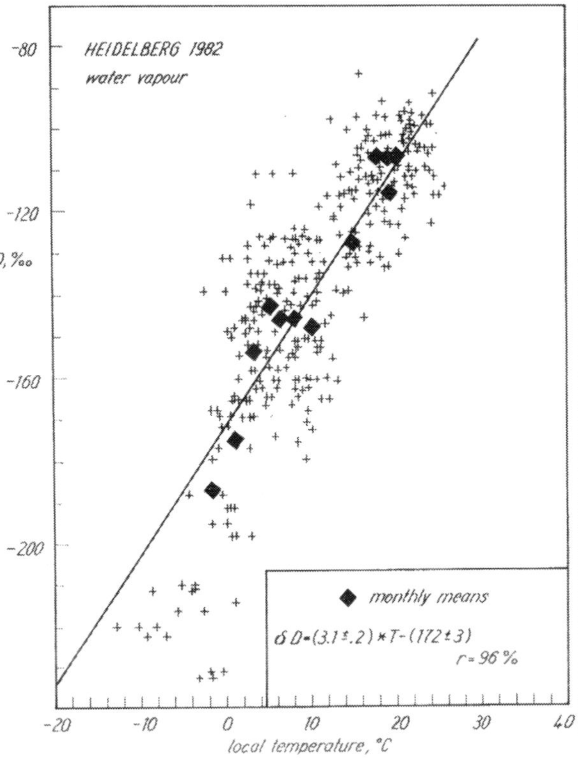

Fig. 5. Temperature correlation of δD in mean daily water vapour at Heidelberg

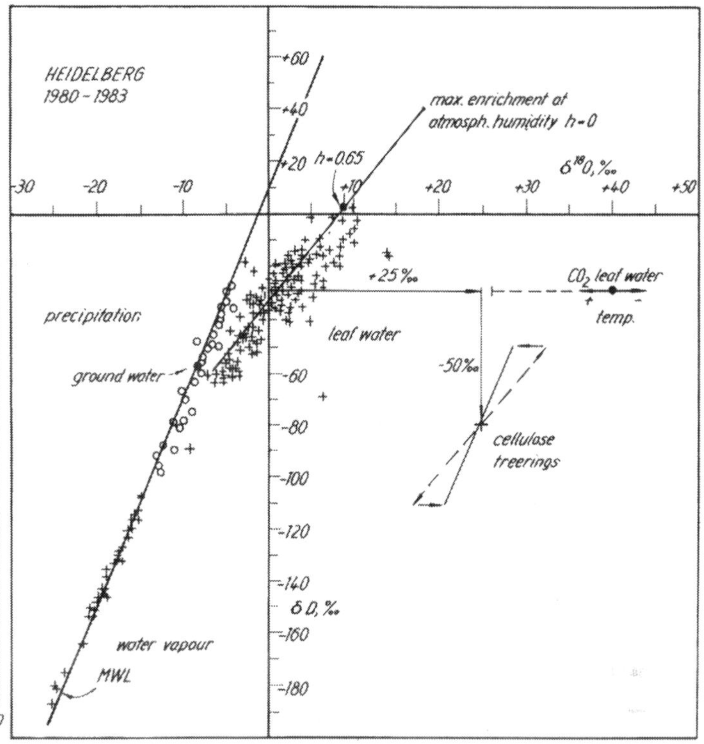

Fig. 6. δD versus δ^{18}O of mean monthly atmospheric water vapour, precipitation and of the water from plant leaves, from which the cellulose data point has been derived with relevant isotope fractionations in use

drift with phase velocities increasing with zonal wind [12]. This might have an interesting paleoclimatic implication: A steeper meridional temperature gradient during an ice-age, for example, causes a stronger atmospheric circulation. Then stronger zonal winds imply higher phase velocities of the Rossby Waves and thus more frequent progresses of polar air masses towards lower latitudes and vice versa, i.e. tropical air masses towards higher latitudes. Fig. 5 shows the temperature dependence of δD in mean daily water vapour at Heidelberg. The fairly wide spread of the individual data points around the linear regression line of the slope $\Delta\delta D/\Delta T = 3{\cdot}1\,‰/°C$ cannot be explained yet satisfactorily. This spread may be due to air moisture from different oceanic source regions, as there is the subtropical Atlantic for tropical marine air masses and the North Atlantic for polar marine air masses. We hope that our synoptic vapour study will eventually yield a better understanding of this temperature dependence.

4. Application of δD and δ^{18}O in Paleoclimatology

Apart from high temperature thermal waters, which are normally enriched in δ^{18}O by oxygen exchange with the rock material, groundwaters reflect the isotopic composition of the local precipitation of their infiltration areas. The isotopic depletion of European ice-age waters as being due to lower temperatures at that climatic period has already been mentioned in section 2. The Saharian groundwaters form another example. The deep groundwaters of the Central and Eastern Sahara have ^{14}C-ages exceeding 20000 years B.P. These plaeo-waters show a zonal variation of δD and δ^{18}O [13], which is similar to the one observed in modern European groundwaters (Fig. 1). From this continental effect local groundwater formation by paleo-winter rains from the western drift has been concluded. One might expect, that it should be possible to derive paleo-temperature data from these isotope data with help of the temperature effect. However, this simple procedure leads to unrealistically low temperatures (about +5 °C only) for the paleo-winter season of the Eastern Sahara, which contradicts current paleoclimatic ideas on Saharian temperatures during the last glaciation. In contrast with this mean annual paleo-temperature data obtained from the noble gas content of Saharian groundwaters are only 1 °C to 2 °C lower than the present temperatures [10]. In order to solve this discrepancy, one may speculate that low temperatures might have existed only during rainy weather conditions caused by more frequent advances of polar air masses (see section 3), whereas relatively high temperatures existed during the rest of the time. Altogether, isotope data of paleowaters is expected to yield more detailed paleoclimatic information, if our present knowledge on the interrelation between the isotopes in precipitation and the parameters of the atmospheric water vapour circulation will have been improved. The main disadvantage of groundwater as a plaeoclimatic indicator lies in its inability to record short-term climatic changes by groundwater dating. From this point of view δD and δ^{18}O in plant material (cellulose of tree rings and of peat), speleothems, lake sediments and soil carbonates seem to be much better suited climatic indicators [14]. However, the problem there is, how reliable conclusions from the isotope data of this material towards the isotopic composition of the plaeo-precipitation may be. The δD versus δ^{18}O-plot in Fig. 6 illustrates this problem for plant material: The data points close to the Meteoric Water Line represent the mean monthly water vapour (Fig. 3) and precipitation at Heidelberg. The other data points at the upper right hand side of the line represent the isotopic composition of water from plant leaves, which have been sampled in parallel with the air moisture program for a while.

The higher heavy isotope content of the leaf water in comparison with local rain and soil water is caused by the isotope fractionation at the transpiration surface. This isotopic enrichment of the leaf water is known to depend on the leaf temperature, the moisture deficit $1 - h$ (h = relative humidity) of the air and on the isotopic composition of the air moisture in the environment of the plant. The enrichment can be estimated, if these microclimatic parameters are known [15]. In addition to this (paleo-) microclimatic aspect, the more general problem is to be faced, that plant material is formed during the growing season only. In this warmer season the heavy isotope content of the atmospheric moisture is fairly high, whereas the tree transpires soil water, which is due to precipitation of the previous seasons. In the case of heavy soils with high water retention and trees with deeply reaching roots the soil water used may range back over several years. A further problem is the isotope fractionation between the leaf water and the plant cellulose being measured. As indicated by the cellulose data point of Fig. 6, the cellulose is depleted in δD by about $-50‰$ in comparison to the leaf water. The enrichment in $\delta^{18}O$ is caused by two effects in the very early stages of photo-synthesis: Isotopic equilibration of CO_2 with the leaf water leads to an enrichment of $\delta^{18}O$ in CO_2 in the order of $+42‰$ at room temperature [16]. This fractionation between CO_2 and H_2O decreases with increasing temperature ($\Delta\delta^{18}O/\Delta T \cong -0.22‰/°C$) and thus counteracts the temperature effect of $\delta^{18}O$ in precipitation. Then two thirds of the oxygen, incorporated during glucose formation, come from this ^{18}O enriched CO_2 and one third comes from the leaf water [17] as one would expect from stoechiometry. Contrary to δD, however, there seems to be no further physiological fractionation. Because of these complications it is, at the present state of knowledge, not possible to draw reliable conclusions on the isotopic composition of the paleo-precipitation from the δD and $\delta^{18}O$ data of plant material. Therefore one can only try to interprete relative isotopic variations in treering series or in vertical profiles of peat deposits assuming that they represent temperature variations in the past.

Similar arguments hold for $\delta^{18}O$ of speleothems, lake sediments and soil deposits. The ^{18}O-content of this carbonate material mainly depends on the $\delta^{18}O$ of the original carbonate solution (lake water, soil water) and on the isotope fractionation between this water and the carbonate precipitate. Equilibrium isotope fractionation at room temperature leads to an enrichment of the calcite precipitate in the order of $+30‰$ [18], which is again temperature dependent ($\Delta\delta^{18}O/\Delta T = -0.24‰/°C$) and counteracting the temperature effect of $\delta^{18}O$ in precipitation. Moreover, an additional kinetic fractionation comes often into play, which cannot be accounted for with sufficient accuracy. Another complication lies in the fact, that the ^{18}O-content of the original solution, which is reflected by the $\delta^{18}O$ of the carbonates, cannot be considered automatically to be identical with the one of the paleo-precipitation. Evaporation effects might have caused an isotopic enrichment of the carbonate solution relative to the local precipitation, which cannot easily be accounted for if δD data are not available.

Acknowledgement

We wish to thank Dr. *L. Merlivat* and Dr. *J. Jouzel*, Centre d'Etudes Nucleaires de Saclay, France, for sampling and measuring water vapour at Paris, as well as Prof. *M. Geyh*, Niedersächsisches Landesamt für Bodenforschung, Hannover, Fed. Rep. of Germany and Prof. *G. Östlund*, University of Miami, School of Marine and Atmospheric Science, Florida, USA, for the water vapour sampling at their institutes.

For excellent cooperation we thank Dr. *K. Rozanski*, University of Mining and Metallurgy, Inst. of Physics and Nuclear Techniques, Cracow, Poland, who provided the isotope data of Cracow.

Special thansk to Mrs. *Ch. Junghans*, who ran the stable isotope measurements.

This investigation was supported by the Heidelberger Akademie der Wissenschaften.

Received July 9, 1984
Accepted in revised form October 4, 1984

References

[1] IAEA, 70, 71, 73, 75 and 1979: Environmental Isotope Data. World survey of isotope concentrations in precipitation. Technical Report Series No. 69, 117, 129, 147, 165, 192, IAEA Vienna 1969

[2] *J. Gat*, The isotopes of hydrogen and oxygen in precipitation. *P. Fritz* & *J. Ch. Fontes* (editors): Handbook of Enivonmental Isotope Geochemistry, Elsevier, Amsterdam – Oxford – New York 1980, The Terrestrial Environment, Vol. 1, 21 – 47

[3] *Y. Yurtsever*, Worldwide survey of stable isotopes in precipitation. Rep. Sect. Isotope Hydrology. IAEA, November 1975, 40 pp.

[4] *E. Eriksson*, Tellus **17** (1965) 498

[5] *C. J. Yapp, S. Epstein*, Earth & Planetary Science Letters **34** (1977) 333

[6] *C. H. Reitan*, Bulletin Am. Meteorol. Society **41** (1960) No. 2, 79

[7] *C. Sonntag, K. Rozanski, K. O. Münnich, H. Jacob*, Variations of deuterium and oxygen 18 in continental precipitation and groundwater, and their causes. *A. Street-Perott* (ed.): Variations in the global water budget. D. Reidel Publishing Company (1983), 107 to 124

[8] *K. Rozanski, C. Sonntag, K. O. Münnich*, Tellus **34** (1982) 142

[9] *W. Dansgaard*, Tellus **16** (1964) 436

[10] *J. Rudolph, H. K. Rath, C. Sonntag*, Noble gases and stable isotopes in ^{14}C-dated paleowaters from Central Europe and the Sahara. Proc. of Int. Symp. on Isotope Hydrology in Water Resources Development, IAEA Vienna, 12 – 16 Sept. 1983 (in press)

[11] *H. Schoch-Fischer, K. Rozanski, H. Jacob, C. Sonntag, J. Jouzel, G. Ostlund, M. A. Geyh*, Hydrometeorological factors controlling the time variation of D, ^{18}O and ^{3}H in atmospheric water vapour and precipitation in the Northern Westwind belt. Proc. of Int. Symp. on Isotope Hydrology in Water Resources Development, IAEA Vienna, 12 – 16 Sept. 1983 (in press)

[12] *G. H. Liljequist, K. Cehak*, Allgemeine Meteorologie. Braunschweig/Wiesbaden: P. Friedr. Vieweg & Sohn 1979

[13] *C. Sonntag, E. Klitzsch, E. P. Löhnert, E. M. El-Shazly, K. O. Münnich, Ch. Junghans, U. Thorweihe, K. Weistroffer, F. M. Swailem*, Paleoclimatic information from deuterium and oxygen 18 in carbon-14 dated North Saharian groundwaters. Isotope Hydrology 1978, Vol. II, 569 – 581, IAEA Vienna

[14] IAEA, Paleoclimated and Paleowaters: A collection of environmental isotope studies. Proc. Advisory Group meeting Vienna, 25 – 28 November 1980, Vienna IAEA 1983

[15] *H. Förstel*, Rad. and Environm. Biophys. **15** (1978) 323

[16] *Y. Horibe, K. Shigehara, Y. Takakuwa*, J. Geophys. Res. **78** (1973) 2625

[17] *De Niro, S. Epstein*, Science **204** (1979) 51

[18] *Y. Bottinga*, J. Phys. Chem. **72** (1968) 800

Isotopenpraxis **21** (1985) 6, S. 199 bis 203

Berechnung von thermodynamischen Isotopieeffekten einfacher Fluide über Quantenkorrekturen und isotopiebedingte Unterschiede in Potentialparametern

H. Schmid

(Zentralinstitut für Isotopen- und Strahlenforschung der Akademie der Wissenschaften der DDR, Leipzig)[1])

Es werden Formeln für thermodynamische Isotopieeffekte einfacher Fluide angegeben. Diese Fluide werden hier als Hartkugel- und Square-Well-Systeme behandelt. Eine spezielle Methode, die Mikropotentialmethode, wird zur Berechnung der klassischen thermodynamischen Funktionen benutzt. Quantenkorrekturen zu diesen Funktionen sind ebenfalls angegeben. Durch Änderung der Gesamtmasse des Moleküls hervorgerufene Isotopieeffekte (Masseneffekte) werden auf die Berechnung der Veränderung der Quantenkorrekturen thermodynamischer Funktionen zurückgeführt. Isotopieffekte infolge Änderung von Potentialparametern (Potentialeffekte) werden für Methan explizit angegeben. Die Potentialeffekte von Methan sind von der gleichen Größenordnung wie die entsprechenden Masseneffekte.

Formulae are given for thermodynamic isotope effects of simple fluids. These fluids are here treated as hard-sphere and square-well systems. A special method, the micropotential method, is used for the calculation of classical thermodynamic functions. Quantum corrections to these functions are also given. Isotope effects due to the variation of the total mass (mass effects) are related to the estimation of quantum corrections of thermodynamic functions. Isotope effects due to the variation of potential parameters (potential effects) are calculated for methane. The potential effects of methane are of the same order of magnitude as the corresponding mass effects.

Keywords

fluids; intermolecular forces; isotope effects; methane; rare gases; thermodynamic properties

1. Einführung

In dieser Arbeit werden einfache Systeme untersucht, die aus zumindest näherungsweise sphärisch symmetrischen Molekülen bestehen. Diese Teilchen wirken dabei durch paarweise additive Zentralkräfte aufeinander. Bei Kenntnis der gesamten potentiellen Energie eines einfachen Systems (z.B. verflüssigte Edelgase und Methan) kann man die Zustandsumme Q und mit Hilfe von Q thermodynamische Funktionen bestimmen (Abschnitt 2). Dabei werden zusätzlich auch Quanteneffekte erfaßt. Mit den Ergebnissen aus Abschnitt 2 werden dann im Abschnitt 3 thermodynamische Isotopieeffekte einfacher (dichter) Fluide, d. h. Veränderungen thermodynamischer Eigenschaften bei Isotopensubstitutionen bestimmt. Solche Effekte ergeben sich aus

$$\Delta f_X = f_X(\text{Isotop 1}) - f_X(\text{Isotop 2}) \qquad (1)$$

wobei f_X eine beliebige thermodynamische Funktion des einkomponentigen Systems ist [1, 2]. Für Neon, Argon und Methan sind nach Gleichung (1) Isotopieeffekte berechnet worden.

Im Abschnitt 4 werden die Ergebnisse diskutiert und Schlußfolgerungen gezogen.

2. Bestimmung thermodynamischer Funktionen nach der Mikropotentialmethode

2.1. *Klassische Resultate*

Mit einer neuen, zur Berechnung von Flüssigkeitseigenschaften bisher nicht benutzten Methode, der Mikropotentialmethode (MPM), wird das Konfigurationsintegral dichter (klassischer) Fluide bestimmt. Die MPM ermöglicht es, mittels gewisser Wahrscheinlichkeitsdichtefunktionen von „Mikropotentialen" und entsprechender erzeugender Funktionen gute Näherungen für das Konfigurationsintegral und damit für die Zustandssumme Q anzugeben [3]. Das „Mikropotential" U ist dabei die potentielle Energie eines (beliebig) herausgegriffenen Teilchens, die durch die Konfiguration der anderen Moleküle bestimmt ist. Zur Angabe des „Mikropotentials" U und des Konfigurationsintegrals ist die Kenntnis des Paarpotentials u notwendig. Als Paarpotential wird das Square-Well-Potential

$$\left.\begin{aligned} u(r) &= \infty \quad \text{für} \quad 0 \leqq r \leqq \sigma \\ &= \varepsilon \quad \text{für} \quad \sigma < r < c\,, \quad \varepsilon < 0 \\ &= 0 \quad \text{für} \quad r \geqq c \end{aligned}\right\} \qquad (2)$$

benutzt, wobei $|\varepsilon|$ die Potentialtopftiefe, σ der Hartkugeldurchmesser und $c = 1{,}5\sigma$ sind. Das Paarpotential von (2) berücksichtigt in einfachster Weise das Wirken abstoßender und anziehender Kräfte. Falls $\varepsilon = 0$ ist, geht (2) in ein Hartkugelpotential über. Mit dem Paarpotential $u(r)$ gemäß (2) erhält man für ein einkomponentiges System nach der MPM in guter Näherung die Zustandssumme

$$Q = \frac{\lambda^{-3N}}{N!} \cdot V^N \cdot \exp\left\{\frac{4\pi\sqrt{2}}{3} N f^* \left[A \ln(1 - D_V) + B\left(\left(\frac{c}{\sigma}\right)^3 - 1\right) \ln D_1\right]\right\} \qquad (3)$$

mit

$$\left.\begin{aligned} &B = \frac{1{,}5^3 - A}{1{,}5^3 - 1}\,, \quad D_V = \frac{V_0}{V}\frac{1}{f^*}\,, \quad D_T = e^{\frac{1}{2T^*}}\,, \\ &D_1 = D_V(D_T - 1) + 1 \end{aligned}\right\} \qquad (4)$$

[1]) Anschrift: DDR-7050 Leipzig, Permoserstraße 15

Dabei sind $A = 0{,}5$, $f^* = 0{,}7950$, $T^* = KT/|\varepsilon|$ die reduzierte Temperatur, $V_0 = N\sigma^3/\sqrt{2}$, und

$$\lambda = \left(\frac{2\pi\hbar^2}{mKT}\right)^{1/2} \tag{5}$$

ist die *de Broglie*-Wellenlänge. Mit Hilfe der freien Energie

$$F = -KT \ln Q \tag{6}$$

und thermodynamischer Relationen kann man aus (3) alle thermodynamischen Funktionen berechnen [4, 5]. Aus (3) und (5) ergeben sich dann insbesondere der um Eins verminderte Kompressibilitätsfaktor

$$\frac{pV}{NKT} - 1 = \frac{4\pi\sqrt{2}}{3} \cdot f^* \cdot \left\{A \cdot \frac{D_V}{1 - D_V} - B \times \left[\left(\frac{c}{\sigma}\right)^3 - 1\right] \frac{D_1 - 1}{D_1}\right\} \tag{7}$$

und die auf das ideale Gas bezogene reduzierte innere Energie

$$\frac{E - E_0}{N|\varepsilon|} = -\frac{1}{2}\frac{4\pi\sqrt{2}}{3} \cdot f^* \left[\left(\frac{c}{\sigma}\right)^3 - 1\right] \cdot B\, \frac{D_V \cdot D_T}{D_1} \tag{8}$$

mit E_0 als innerer Energie des idealen Gases.

2.2. *Quantenkorrekturen*

Thermodynamische Isotopieeffekte beruhen auf dem Quantencharakter der Teilchen [1, 2]. Eine Berechnung der quantenstatistischen Zustandssumme ist für (dichte) Fluide allgemein nicht durchführbar. Nur vereinfachte Modelle lassen sich voll quantenstatistisch behandeln (siehe z. B. [6]). Auf solche exakten Berechnungen kann verzichtet werden, wenn die Quanteneffekte als (kleine) Korrekturen zu den klassischen Resultaten betrachtet werden können. Die klassische Methode zur störungstheoretischen Berücksichtigung von Quantenkorrekturen ist die bekannte *Wigner-Kirkwood*-Entwicklung [7, 8]. Eine wesentliche Beschränkung der Anwendbarkeit derselben ist dadurch gegeben, daß die Berechnung der Korrekturterme die Differentiation des Paarpotentials u erfordert. Die in dieser Arbeit verwendeten Square-Well- und Hartkugelpotentiale (Gleichung (2) bzw. (2) mit $\varepsilon = 0$) erfüllen das Kriterium für die Anwendbarkeit der *Wigner-Kirkwood*-Methode nicht. Allgemein erhält man für Paarpotentiale, die nichtanalytische Funktion des Abstandes der Moleküle sind nach *Gibson* [9] die Zustandssumme semiklassischer Fluide

$$Q^{\text{korr}} = Q + Q_1 \cdot \lambda + Q_2 \cdot \lambda^2 + \ldots \tag{9}$$

mit Q als klassischer Zustandssumme und λ gemäß Gl. (5) Bei der Entwicklung (9) wurde die quantisierte *Boltzmann*statistik zugrunde gelegt.

Von besonderer praktischer Bedeutung ist indes eine spezielle, physikalisch plausible Methode, um Quantenkorrekturen 1. Ordnung in λ zu berechnen (siehe [9–12]). Dieses als Methode des effektiven Durchmessers bekannte Verfahren geht davon aus, daß im semiklassischen Fall (kleine λ) die Quantennatur des Hartkugelsystems exakt durch die Ortsunschärfe der Teilchen insofern bestimmt ist, daß man bei ansonsten klassischem Verhalten den Teilchendurchmesser σ in bestimmter Weise vergrößert. Dabei gilt $\sigma \to \sigma + 2^{-3/2} \cdot \lambda$. Für eine beliebige thermodynamische Größe f_x (einschließlich der Zustandssumme Q nach (3)) folgt dann bei einer Entwicklung nach λ bis zur 1. Ordnung

$$f_x^{\text{quant}} = \left(1 + 2^{-3/2} \cdot \lambda \frac{\partial}{\partial\sigma}\right) f_x^{\text{klass}} \tag{10}$$

wobei f_x die freie Energie, der Druck usw. sein kann und λ durch (5) gegeben ist.

Mit den Ergebnissen der MPM erhält man nach (10) für ein Hartkugelsystem die Druckkorrektur Δp 1. Ordnung durch

$$\frac{\Delta p}{\varrho KT} = \frac{p^{\text{quant}} - p^{\text{klass}}}{\varrho KT} = \frac{\lambda}{\sigma}\left[2\pi A f^* \frac{D_V}{(1 - D_V)^2}\right] \tag{11}$$

wobei p^{klass} durch die Gleichung (7) bestimmt ist, wenn dort $\varepsilon = 0$ gesetzt wird, und es ist $\varrho = N/V$. Die Beziehung (11) zeigt, daß stets $p^{\text{quant}} > p^{\text{klass}}$ ist, was wegen der Ortsunschärfe der Teilchen zu erwarten war. In Abb. 1 ist $p^{\text{quant}}/(\varrho KT)$ für verschiedene λ dargestellt und wird mit dem klassischen Resultat ($\lambda = 0$) verglichen.

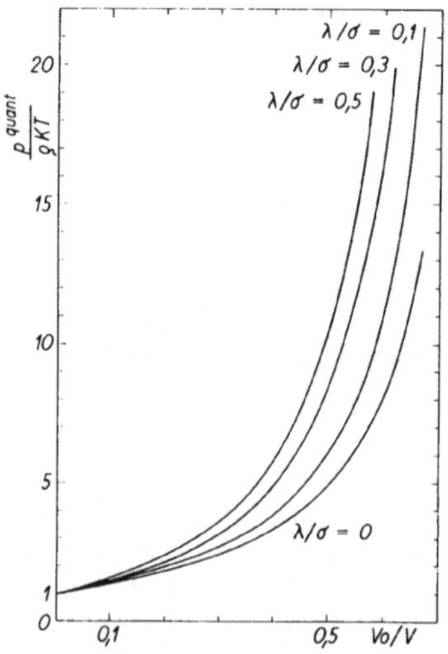

Abb. 1. Kompressibilitätsfaktor mit Quantenkorrektur für ein Hartkugelsystem

Wählt man als Paarpotential ein Square-Well-Potential und benutzt die Beziehung (10), so ergibt sich

$$\frac{\Delta p}{\varrho KT} = \frac{\lambda}{\sigma} \cdot 2\pi A f^* \frac{D_V}{(1 - D_V)^2} + \frac{\lambda}{\sigma} \cdot 2\pi \left[\left(\frac{c}{\sigma}\right)^3 - 1\right] B f^* \frac{1 - D_1}{D_1^2} \tag{12}$$

Der zweite Summand auf der rechten Seite von (12) entsteht dabei durch zusätzliche Berücksichtigung anziehender Kräfte und bewirkt im Gegensatz zum Hartkugelterm (1. Summand) eine Druckverminderung, was beim Square-Well-System infolge der Ortsunschärfe durchaus vernünftig ist. Es zeigt sich, daß der Betrag des zweiten Summanden in (12) maximal wird, wenn

$$\frac{V_0}{V} \cdot \frac{1}{f^*}\left(e^{\frac{|\varepsilon|}{2KT}} - 1\right) = 1 \tag{13}$$

gilt, und dabei den Wert $(\lambda/\sigma) \cdot 3{,}590$ annimmt, so daß bei höheren Dichten der erste Summand in (12) den entscheidenden Quanteneffekt liefert. Bei niedrigen Dichten ($V_0/V \ll 1$) ändert der Ausdruck (12) jedoch das Vorzeichen, wenn $T^* = KT/|\varepsilon| < 3{,}118$ ist.

Zusätzlich müssen aber bei Entwicklungsgliedern höherer Ordnung als (λ/σ) noch Terme mit $\Delta\varepsilon^*$ ($|\varepsilon| \to |\varepsilon| - |\Delta\varepsilon^*|$) wegen der Ortsunschärfe berücksichtigt werden. Für $|\Delta\varepsilon^*|$ kann man als genäherten Wert z.B. die Höhe des 1. Energieniveaus beim unendlich hohen kugelsymmetrischen Potentialtopf (vgl. [13]) angeben. Es wird dann

$$\frac{|\Delta\varepsilon^*|}{2KT} \sim 2\pi\left(\frac{\lambda}{\sigma}\right)^2 \tag{14}$$

mit λ aus Beziehung (5). Dabei verringert sich $|\Delta\varepsilon^*|$ noch, wenn Potentialtöpfe mit endlicher Tiefe zur Modellierung des Potentials benutzt werden, und die Entwicklungsglieder in $\left(\frac{\Delta\varepsilon^*}{2KT}\right)$ sind von 2. Ordnung in (λ/σ).

3. Isotopieeffekte

Für einatomige und eine Reihe kleinerer mehratomiger Moleküle wird der wesentliche Beitrag zum Isotopieeffekt durch die Änderung der Gesamtmasse der Moleküle bei Isotopensubstitution (Masseneffekt) bestimmt [14]. Innere Freiheitsgrade spielen nur eine geringe Rolle.

Für Hartkugelsysteme (es gilt Gl. (2) mit $\varepsilon = 0$) erhält man nach Beziehung (9) Quantenkorrekturen. Speziell für die Bestimmung der Korrekturen 1. Ordnung eines Hartkugelsystems wird Gleichung (10) benutzt. Diese führt bei Verwendung der klassischen Hartkugelresultate ($\varepsilon = 0$) nach Gl. (7) zur Beziehung (11) für die Druckkorrektur Δp. Damit ergibt sich für den Isotopieeffekt des Kompressibilitätsfaktors

$$\frac{\Delta p^{\text{quant}}}{\varrho KT} = \frac{p^{\text{quant}}(m+\Delta m)}{\varrho KT} - \frac{p^{\text{quant}}(m)}{\varrho KT}$$
$$= [\lambda(m+\Delta m) - \lambda(m)]\frac{1}{\sigma}\left[2\pi f^* \frac{D_V}{(1-D_V)^2}\right] \tag{15}$$

wobei

$$\Delta\lambda = \lambda(m+\Delta m) - \lambda(m) = \lambda(m)\cdot\left(\sqrt{\frac{m}{m+\Delta m}} - 1\right) \tag{16}$$

ist. Für Neon, Methan und Argon sind nach der Beziehung (15) Isotopieeffekte bestimmt worden. In Tab. 1 sind $\frac{p^{\text{quant}}(m)}{\varrho KT}$ und $\frac{\Delta p^{\text{quant}}}{\varrho KT}$ für Neon angegeben. Dabei werden ^{20}Ne und ^{22}Ne bei $T = 30$ K und $T = 150$ K verglichen. Es zeigt sich – wie erwartet –, daß die isotopiebedingte Änderung des Kompressibilitätsfaktors mit steigender Temperatur abnimmt, und daß die $p/(\varrho KT)$-Werte für das schwere Isotop ^{22}Ne bei gleicher Dichte kleiner sind als die entsprechenden Größen für ^{20}Ne. Analoges Verhalten zeigen Methan und Argon. Für Methan ist in Abb. 2 $\frac{\Delta p^{\text{quant}}}{\varrho KT}$ für $T = 100$ K und $T = 200$ K in Abhängigkeit von der reduzierten Dichte angegeben.

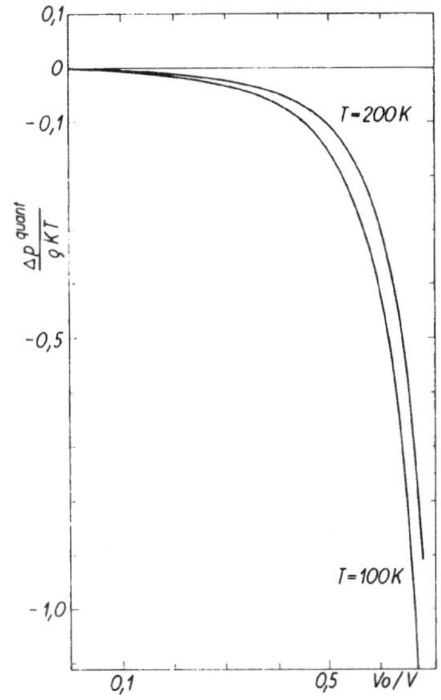

Abb. 2. Isotopieeffekt (Masseneffekt) für Methan als Hartkugelsystem

Bei mehratomigen Molekülen kann die Isotopensubstitution zu Veränderung von Größe, Gestalt und Polarisierbarkeit der Moleküle führen. Insbesondere bei Deuterierung ergeben sich merkbare Veränderungen der zwischenmolekularen Kräfte. Während beim Neon und Argon solche Effekte naturgemäß nicht zu erwarten sind, werden sie jedoch z.B. beim Methan wesentlich [14]. Es verkleinert sich beim Methan infolge Deuterierung die Molekülgröße um 4‰ ($\Delta\sigma/\sigma = -0{,}004$), während sich die Bindungswinkel (die Gestalt) der Moleküle kaum ändern. Die Verkleinerung des Moleküldurchmessers bei der Deuterierung des Methans beruht auf einer Verkürzung der Bindungslänge der CD-Bindungen gegenüber den CH-Bindungen. Theoretisch verständlich wird dieser Effekt (nach *Kuchitsu* und *Bartell* [15]) aus der Anharmonizität des intramolekularen Potentials, die zu einer Verkleinerung der mittleren Schwingungsamplitude der CD-Schwingungen gegenüber den CH-Schwingungen führt, wodurch der mittlere Molekülradius von CD_4 im Vergleich zu CH_4 verringert wird. Es ergibt sich deshalb für Methan (als

Tab. 1. Isotopieffekte (Masseneffekte) von Neon als Hartkugelsystem

V_0/V	$T = 30$ K		$T = 150$ K	
	$\frac{p^{\text{quant}}(m)}{\varrho KT}$	$\frac{\Delta p^{\text{quant}}}{\varrho KT}$	$\frac{p^{\text{quant}}(m)}{\varrho KT}$	$\frac{\Delta p^{\text{quant}}}{\varrho KT}$
0,0750	1,33098	−0,00399	1,37300	−0,00178
0,1538	1,78664	−0,0103	1,66413	−0,00461
0,2500	2,5785	−0,0232	2,3031	−0,0104
0,3536	3,96025	−0,0500	3,24040	−0,0224
0,4243	5,52263	−0,0851	4,5122	−0,0381
0,4950	8,1410	−0,152	6,3410	−0,0678
0,5657	13,179	−0,297	9,658	−0,133
0,6250	22,464	−0,596	15,3840	−0,267
0,6670	37,378	−1,122	24,0520	−0,502

Tab. 2. Isotopieeffekte (Potentialeffekte) von Methan als Square-Well-System

V_0/V	$T^* = 0{,}707$		$T^* = 1$		$T^* \gg 1$	
	$p/(\varrho KT)$	$\frac{p + \Delta p_{ges}^{Pot}}{\varrho KT}$	$p/(\varrho KT)$	$\frac{p + \Delta p_{ges}^{Pot}}{\varrho KT}$	$p/(\varrho KT)$	$\frac{p + \Delta p_{ges}^{Pot}}{\varrho KT}$
0,0750	0,0479	0,7013	0,4644	0,4783	1,24528	1,24205
0,1538	−0,8622	−0,6448	0,0549	0,0778	1,56502	1,55667
0,2500	−1,228	−1,1859	−0,2140	−0,1883	2,0802	2,0614
0,3536	−1,3630	−1,33185	−0,1457	−0,1316	2,8859	2,8457
0,4243	−1,10284	−1,09499	0,21273	0,20504	3,6947	3,6266
0,4950	−0,4003	−0,4418	0,9894	0,9337	4,885	4,764
0,5657	1,0878	0,9351	2,533	2,367	6,809	6,575
0,6250	3,604	3,223	5,084	5,692	9,657	9,194
0,6670	6,999	6,225	8,498	7,713	13,270	12,413

Hartkugelsystem betrachtet) ein zusätzlicher Isotopieeffekt

$$\frac{\Delta p^{Pot}}{\varrho KT} = \frac{4\pi\sqrt{2}}{3}\cdot A \cdot f^*\left\{\frac{D_V^*}{1-D_V^*} - \frac{D_V}{1-D_V}\right\} \qquad (17)$$

mit

$$D_V^* = D_V\left(1 + \frac{\Delta\sigma}{\sigma}\right)^3 \qquad (18)$$

Vergleicht man diesen mit dem reinen Masseneffekt nach Gl. (15) für $T = 100$ K, so sind beide Isotopieeffekte gleichgerichtet und liegen in der gleichen Größenordnung. Ein analoges Bild ergibt sich für $T = 200$ K, wobei hier zwar $\frac{\Delta p^{quant}}{\varrho KT}$ betragsmäßig kleiner als $\frac{\Delta p^{Pot}}{\varrho KT}$ ist, jedoch in der gleichen Größenordnung liegt. In beiden Fällen vergrößert sich $|\Delta p^{Pot}/(\varrho KT)|$, wenn die Dichte anwächst.

Betrachtet man im Methan auch die anziehenden Kräfte zwischen den Molekülen, so ist neben dem Effekt nach Gl. (17) noch ein weiterer Potentialeffekt zu berücksichtigen. Wie in [14] angegeben wird, verändert sich wegen der Differenz der mittleren Bindungslänge von C—H und C—D bei Deuterierung des Methans auch die Polarisierbarkeit der Moleküle, was zu einer Verkleinerung der Tiefe ε des Paarpotentials um $9‰$ ($\Delta\varepsilon/\varepsilon = -0{,}009$) führt. Der Wert von c/σ bleibt unverändert, d. h., $c/\sigma = 1{,}5$. Vernachlässigt man beim Methan Masseneffekte und behandelt es als ein durch (2) definiertes Square-Well-System, so können mit Hilfe der Beziehungen (7) und (8) Potentialeffekte für Methan berechnet werden. In Tab. 2 sind $p/(\varrho KT)$ und $p + \Delta p_{ges}^{Pot}$ bei verschiedenen reduzierten Temperaturen ($T^* = 0{,}707$ entspricht für Methan 100 K) in Abhängigkeit von der reduzierten Dichte V_0/V angegeben, wobei p der Druck nach Gl. (7) und Δp_{ges}^{Pot} der gesamte Potentialeffekt sind. Es gilt

$$\frac{p + \Delta p_{ges}^{Pot}}{\varrho KT} = 1 + \frac{4\pi\sqrt{2}}{3}\cdot f^*\left\{A\,\frac{D_V^*}{1-D_V^*} + B \times \left[\left(\frac{c}{\sigma}\right)^3 - 1\right]\frac{1-D_1^*}{D_1^*}\right\} \qquad (19)$$

mit

$$D_T^* = D_T \cdot e^{\frac{|\Delta\varepsilon|}{2KT}}, \qquad D_1^* = D_V^*(D_T^* - 1) + 1 \qquad (20)$$

Erwartungsgemäß verringert sich dabei infolge der Deuterierung der Druck bei hohen Dichten, während bei kleiner werdenden Dichten schließlich der Einfluß von $\Delta\varepsilon$ überwiegt, wodurch der Druck vergrößert wird ($\Delta p_{ges}^{Pot} > 0$). $\Delta p_{ges}^{Pot}/(\varrho KT)$ selbst ist in Abhängigkeit von der reduzierten

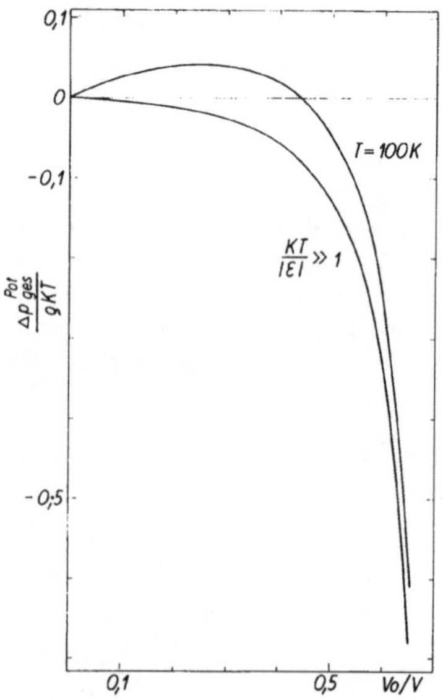

Abb. 3. Isotopieeffekt (Potentialeffekt) für Methan als Square-Well-System

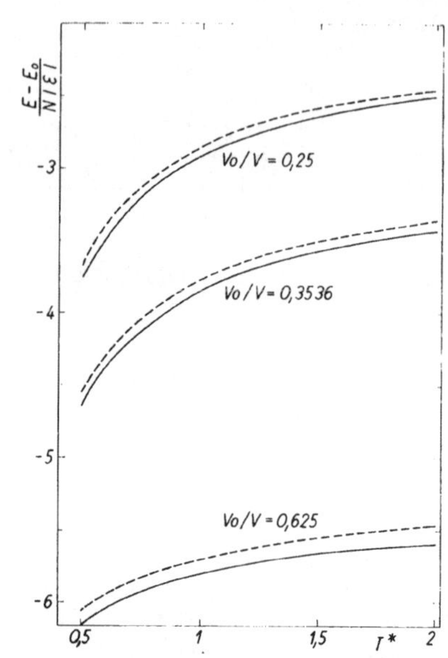

Abb. 4. Reduzierte, auf das ideale Gas bezogene innere Energie von Methan als Square-Well-System
——— Methan; – – – deuteriertes Methan

Dichte V_0/V für $T = 100$ K und $KT/|\varepsilon| \gg 1$ in Abb. 3 dargestellt. Für geringe und mittlere Dichten ergibt sich ein inverser Isotopieeffekt, falls $T^* < 5{,}379$ ist.

Abb. 4 zeigt die generelle und erwartete Zunahme der inneren Energie infolge Isotopensubstitution. Dabei ist $\frac{E - E_0}{N|\varepsilon|}$ der Ausdruck nach Gl. (8), und es gilt

$$\frac{(E + \Delta E) - E_0}{N|\varepsilon|} = -\frac{1}{2}\frac{4\pi\sqrt{2}}{3}\cdot f^* \times B\left[\left(\frac{c}{\sigma}\right)^3 - 1\right]\frac{D_T^* \cdot D_V^*}{D_1^*} \quad (21)$$

wobei E_0 die innere Energie des idealen Gases ist. In Abhängigkeit von der reduzierten Temperatur werden in Abb. 4 die reduzierten Exzessgrößen der inneren Energie bei verschiedenen Dichten dargestellt. Mit zunehmender Dichte nimmt dabei ΔE zu.

Abschließend sei noch auf den Masseneffekt eines Square-Well-Systems eingegangen. Da die Beziehungen (12) und (14) einer genaueren Darstellung und Begründung bedürfen, soll dies nur kurz geschehen. Aus der Gl. (12) folgt

$$\frac{p^{\text{quant}}(m + \Delta m)}{\varrho KT} - \frac{p^{\text{quant}}(m)}{\varrho KT} = \frac{\lambda(m)}{\sigma}\cdot\left(\sqrt{\frac{m}{m + \Delta m}} - 1\right) \times\left\{2\pi A f^* \frac{D_V}{(1 - D_V)^2} - 2\pi(1{,}5^3 - A)\, f^* \frac{D_1 - 1}{D_1^2}\right\} \quad (22)$$

Analog zur Gl. (12) und den sich dort anschließenden Betrachtungen erhält man, daß der 1. Term in der geschweiften Klammer von (22) bei hohen Dichten im fluiden Bereich um mindestens eine Größenordnung größer als der 2. Term in dieser Klammer ist. Beim Übergang zu niedrigen Dichten bis hin zu $V_0/V \ll 1$ ändern jedoch die Ausdrücke (12) und (22) das Vorzeichen, wenn $T^* \leq 3{,}118$ gilt. Es ist damit für den Ausdruck von (22) ein ähnliches Verhalten wie für $\Delta p_{\text{ges}}^{\text{Pot}}$ aus der Beziehung (19) zu erwarten, wenn die Dichte bei gegebener Temperatur sich ändert. Der Einfluß anziehender Kräfte ($\varepsilon < 0$ in Beziehung (2)) führt nach Gl. (22) zu einer Vergrößerung des Druckes bei Isotopensubstitution.

4. Diskussion und Schlußfolgerungen

Die Mikropotentialmethode (MPM) liefert für die thermodynamischen Funktionen eines Square-Well-Systems Ergebnisse, die mit denen von entsprechenden Computersimulationen recht gut übereinstimmen (siehe [3] und dort zitierte Literatur). Als vorteilhaft erweist sich dabei, daß die MPM für das Konfigurationsintegral und die thermodynamischen Funktionen analytische Ausdrücke liefert. Dies ermöglicht in einfacher Weise die Berechnung von Quantenkorrekturen und thermodynamischen Isotopieeffekten.

Für Square-Well- und Hartkugelsysteme erhält man nach *Gibson* [9] eine Entwicklung der Zustandssumme nach geraden und ungeraden Potenzen der *de Broglie*-Wellenlänge λ. Speziell bei der Berechnung der Quantenkorrekturen thermodynamischer Funktionen nach Gl. (10) wird die Kenntnis der klassischen Paarverteilungsfunktion nicht benötigt, was normalerweise der Fall ist und die Rechnungen kompliziert. Man weiß aber, daß der Einfluß von Quanteneffekten bei sehr hohen Dichten im fluiden Bereich stark anwächst (vgl. auch Abb. 1), so daß dann zumindest die Gültigkeit einer Entwicklung der Zustandssumme bis zum linearen Glied in λ versagt. Es ist aber schwierig, den Gültigkeitsbereich der Quantenkorrekturen genau festzulegen, da exakte quantenstatistische Vergleichsdaten (etwa aus Simulationsrechnungen) bisher kaum vorliegen. Eine Mögichkeit, Gültigkeitsbereich und Genauigkeit der Korrekturen abzuschätzen, besteht in der Betrachtung einfacher Modelle von (dichten) Fluiden, die quantenstatistisch exakt behandelbar sind (z.B. Zellmodell nach [6]). Die Gl. (11) und (12) führen zu qualitativ vernünftigen Ergebnissen für die Quantenkorrekturen. Insbesondere überwiegt der Hartkugelanteil der Quantenkorrekturen in Gl. (12) bei höheren Dichten.

Unter Benutzung von Gl. (11) und (12) sind nach den Beziehungen (15) und (22) Masseneffekte für Neon, Argon und Methan berechnet bzw. abgeschätzt worden. Die Abschätzung der Masseneffekte für ein Square-Well-System ergibt, daß bei höheren Dichten der Hartkugelanteil deutlich überwiegt, während für niedrigere Dichten ein inverser Isotopieeffekt ($\Delta p^{\text{quant}} > 0$ in Gl. (22)) vorliegt. Es sei hier aber vermerkt, daß die Anwendbarkeit der Beziehungen (12) und (22) im Rahmen einer strengeren Theorie untersucht werden müßte. So ist es nötig, für steigende Dichten und sinkende Temperaturen höhere Entwicklungsglieder in λ aus Gl. (9) zu berücksichtigen, um Quantenkorrekturen genau zu bestimmen.

Beim Methan erfolgt parallel zu den oben diskutierten Masseneffekten auch eine Bestimmung von Potentialeffekten für ein Square-Well-System. Diese Isotopieeffekte infolge Deuterierung führen analog den Masseneffekten eines Square-Well-Systems bei niedrigen Dichten im fluiden Bereich für den Druck zu einem inversen Isotopieeffekt, während bei höheren Dichten das deuterierte Methan einen geringeren Druck als das normale Methan besitzt (vgl. Tab. 2). Betrachtet man nur den Effekt nach Gl. (17) von Methan (als Hartkugelsystem), so sind dieser und der reine Masseneffekt (Gl. (15)) gleichgerichtet und liegen in der gleichen Größenordnung.

Der Autor dankt den Herren Dr. *Heybey* und Dr. *Vörtler* für nützliche Diskussionen und für die stete Unterstützung bei der Durchführung dieser Arbeit.

Eingegangen am 18. 7. 1984

Literatur

[1] *G. Vojta*, Thermodynamische Isotopieeffekte. In: *G. Hertz*, Lehrbuch der Kernphysik, Band III. Leipzig: B. G. Teubner Verlagsgesellschaft 1962, S. 88
[2] *R. Haberlandt*, Isotopenpraxis **12** (1976) 6
[3] *H. Schmid*, zur Veröffentlichung vorgesehen
[4] *J. S. Rowlinson*, Liquids and liquid mixtures. London: Butterworth Scientific Publications 1959, 360 p.
[5] *I. N. Godnew*, Berechnung thermodynamischer Funktionen aus Molekülduten. Berlin: VEB Deutscher Verlag der Wissenschaften 1963, 351 S.
[6] *H.-L. Vörtler*, Dissertation A, AdW Leipzig 1977
[7] *E. Wigner*, Phys. Rev. **40** (1932) 749
[8] *J. G. Kirkwood*, Phys. Rev. **44** (1933) 31
[9] *W. G. Gibson*, Mol. Phys. **30** (1975), 1, 13
[10] *E. A. Mason, T. H. Spurling*, The virial equation of state. Oxford: Pergamon Press 1969, 297 p.
[11] *P. C. Hemmer*, Phys. Lett. **27A** (1968) 377
[12] *B. Jancovici*, Phys. Rev. **178** (1969) 295
[13] *L. D. Landau, E. M. Lifschitz*, Quantenmechanik, 3. Aufl., Berlin: Akademie-Verlag 1967, 638 S.
[14] *A. F. Grigor, W. A. Steele*, J. Chem. Phys. **48** (1968), 1032, 1038
[15] *K. Kuchitsu, L. S. Bartell*, J. Chem. Phys. **36** (1962) 2460, 2470

Isotopenpraxis **21** (1985) 6, S. 204 bis 208

Boron Isotope Separation by Ion Exchange Chromatography Using Weakly Basic Anion Exchange Resin

*S. Itoh, M. Aida, M. Okamoto**), *M. Nomura, Y. Fujii*

(Research Laboratory for Nuclear Reactors Tokyo Institute of Technology)[1])

M. Maeda

(Department of Applied Chemistry, Nagoya Institute of Technology)[2])

Influences of operating temperatures and concentrations of feed boric acid solutions were examined on the above titled process over the ranges of 25 – 70 °C and 0·1 – 1·6 mol/dm³ (M), respectively. The ideal displacement chromatography with a very sharp-cut boundary of a boric acid adsorption band was realized at higher temperatures and lower boric acid concentrations within the experimental conditions. The isotope separation coefficient ε was found to decrease with increases in either temperature or the boric acid concentration. The observed values of ε at 25 °C were 0·013, 0·012 and 0·011 depending on the feed boric acid concentrations of 0·1 M, 0·4 M and 0·8 M, respectively. The ε's at 70 °C were determined as 0·0097 (0·1 M), 0·0086 (0·4 M), 0·0083 (0·8 M) and 0·0073 (1·6 M). The operational condition of temperature 40 °C and 0·4 M of the boric acid concentration was proposed as the optimum for the production of enriched ^{10}B isotope by using the present process.

Der Einfluß von Betriebstemperatur und Konzentration von Borsäurelösungen auf den obengenannten Prozeß wurde im Bereich von 25 ... 70 °C bzw. von 0,1 ... 1,6 mol/dm³ (M) untersucht. Eine ideale Umlagerungschromatographie mit sehr scharfen Grenzen des Borsäureadsorptionsbandes wurde bei höheren Temperaturen und geringerer Borsäurekonzentration unter Versuchsbedingungen festgestellt. Es wurde eine Senkung des Isotopentrennfaktors ε bei Erhöhung sowohl der Temperatur als auch der Borsäurekonzentration festgestellt. Die beobachteten ε-Werte betrugen bei 25 °C, 0,013, 0,012 bzw. 0,011 bei Borsäurekonzentrationen von 0,1 M, 0,4 M bzw. 0,8 M. Die ε-Werte bei 70 °C wurden mit 0,0097 (0,1 M), 0,0086 (0,4 M), 0,0083 (0,8 M) bzw. 0,0073 (1,6 M) ermittelt. Es wurden Betriebsbedingungen mit einer Temperatur von 40 °C und einer Borsäurekonzentration von 0,4 M als optimal für die Produktion des angereicherten ^{10}B-Isotops mit Hilfe des vorliegenden Prozesses vorgeschlagen.

Keywords

boron 10; boron isotopes; chromatography; ion exchange; isotope separation; mass spectroscopy

Introduction

The boron isotope separation by means of strongly basic anion exchange resins has been studied by *Makishima* [1], *Rosset* [2], *Urgell* [3] and *Schütze* [4]. In the previous papers [5 – 7] *Kakihana* reported that the boron isotope separation was achieved by using also weakly basic anion exchange resins. The latter method has a benefit of the simplified operation in the chromatographic process: boron is fed into the ion exchange column in the form of a boric acid solution and eluted out simply by pure (distilled or deionized) water. In this method the ion exchange resin is reloaded with boric acid without any regeneration operations. The former system with strongly basic resins has a relatively large separation factors of ca. 1·02 at ca. 0·1 M of boric acid concentration in solution, but has a disadvantage of the evolution of carbon dioxide gas while in elution. The weakly basic resin system, on the other hand, can completely avoid the CO_2 gas evolution because this system employs neither strongly basic regeneration reagents which usually contain carbonate ions, nor strongly acidic elution reagents which decompose carbonate ions adsorbed in the resins to CO_2 gas.

In the previous work [7] using weakly basic resins, long distance chromatographic experiments were carried out up to 256 m, resulting in the enrichment of ^{10}B isotope to 91·00% from the natural value of 19·84%. In the present work, operational conditions have been examined in order to find the optimum conditions for production of enriched ^{10}B isotope by the concerned process.

Experimental

Materials. The ion exchange resin used was Diaion WA-21 (weakly basic, high-porosity type anion exchange resin, 60 – 80 mesh). The structure and the functional groups of the resin are presented in the previous paper [6]. Other chemicals used were of analytical grade and employed without further purifications.

Apparatus. A set of the appratus for the chromatographic experiments is illustrated in Fig. 1. The column (110 cm high and 2·1 cm ID) was a glass pressure tube with a water jacket to control the operating temperature and packed with the anion exchange resin to the height of 100 cm. A pair of electrodes were inserted at the bottom of the column to monitor the movement of the boron band by measuring the electric conductivity in the resin bed. An injector was stuck through the top rubber stopper to maintain the level of the solution in the column con-

Addresses: [1]) O-okayama 2-12-1, Meguro-ku, Tokyo 152 Japan; [2]) Gokiso-cho, Showa-ku, Nagoya 466, Japan
*) Address for correspondence

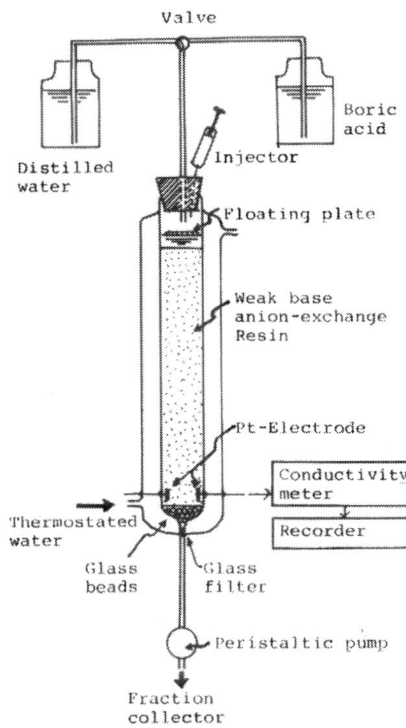

Fig. 1. A set of the apparatus for experiments

stantly at a few mm above the surface of the resin bed. A styro-foam plate was put on the solution in the top of the column in order to prevent the eluent from dropping directly into the resin bed.

Pre-treatment. The anion exchange resin WA-21 was at first converted to the chloride ion form in a batch-wise treatment. Then the resin was uniformly packed into the column and washed with 1 mol/dm³ (M) HCl to remove the impureities in the resin, such as metal and carbonate ions. Finally the resin was again converted to the free-base form with 0·5 M NaOH, and washed thoroughly with distilled water.

Adsorption and Elution. A boric acid solution with a given concentration was fed into the column from a reservoir connected to the column with a polyvinyl chloride tube. The flow rate of the solution was controlled by a peristaltic pump placed beneath the column as shown in Fig. 1. The boric acid solution was charged into the column until the whole resin in the column was completely saturated with boric acid and the isotopic exchange equilibrium was established between the resin and the solution phases.

The boric acid adsorbed in the resin was then eluted with distilled water charged from a top reservoir. The effluent from the column was collected in fractions by a fraction collector. The chromatographic operations were carried out under the experimental conditions of temperatures, 25, 40, 55 and 70 °C, and the concentrations of the boric acid feed solution for adsorption, 0·1, 0·4, 0·8 and 1·6 M. The superficial velocity of the eluent was constant set at 20 cm h^{-1} commonly in each experiment.

Analysis. The concentration of boric acid in each fraction was determined by titration with a standardized NaOH solution after addition of mannitol and drops of phenolphtalein indicator. Te isotopic analysis of boric acid in each sampled fraction was done with a Variain Mat CH-5 mass spectrometer by means of the surface ionization method. The details of the isotope analysis were reported in the previous paper [8].

Results and Discussion

Profile of Chromatogram

The chromatograms for one cycle operation of adsorption and elution of boric acid are illustrated in Fig. 2, where the concentration of boric acid in effluent is shown for the experiments at 40 °C, as examples. Boric acid is fed in a breakthrough manner until the resin reaches the ion exchange equilibrium with the feed solution. The equilibrium point is marked with an arrow in the chromatogram. After feeding additional boric acid to ensure the completion of the isotopic equilibrium between resin and solution phases, the adsorbed boric acid is eluted out by distilled water. The concentration of boric acid in eluent becomes the same values as that of the feed boric acid solution, according to the ion exchange equilibrium. The rear boundaries of the boric acid bands depend on the initial loading condition: sharp-cut boundaries are realized in cases of the low-concentration feed boric acid, e.g. 0·1 M and of the higher temperatures. The shape of the chromatograms suggests that the ion exchange selectivity for boric acid

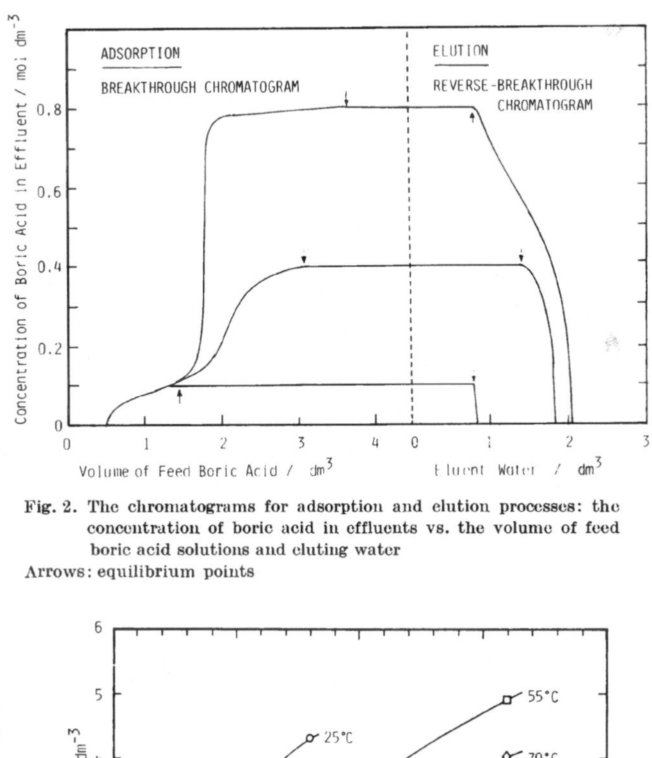

Fig. 2. The chromatograms for adsorption and elution processes: the concentration of boric acid in effluents vs. the volume of feed boric acid solutions and eluting water
Arrows: equilibrium points

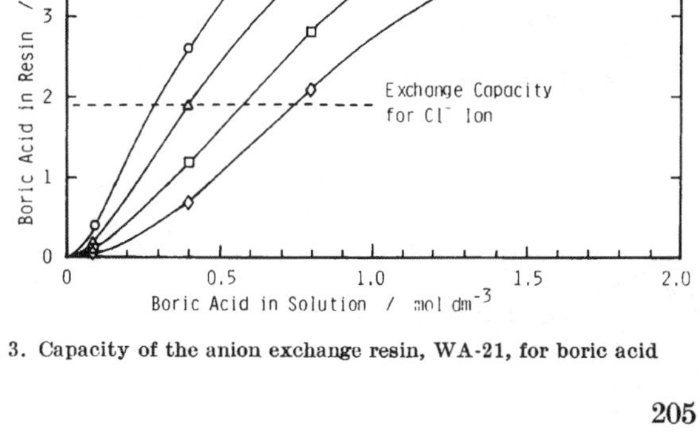

Fig. 3. Capacity of the anion exchange resin, WA-21, for boric acid

is complicated in relation to the concentration of boric acid in solution.

The integration of the elution chromatogram in Fig. 2 gives the total amount of adsorbed boric acid in the resin being equilibrated with a feed solution. Making a correction on void fractions in the column, the adsorption capacity of the resin is calculated and plotted in Fig. 3 for each experiment. The curves in Fig. 3 represent the ion exchange isotherms for the system of boric acid and WA-21 anion exchange isotherms for the system of boric acid and WA-21 exchange resin. The standard exchange capacity of this resin for Cl^- ion is indicated by the dashed line.

It is pointed out in Fig. 3 that (1) the adsorption capacity for boric acid exceeds the normal capacity of the resin when contacted with a concentrated boric acid solution, and (2) the isotherm shows an S-shape with a gentle slope at both low concentration and very high concentration regions. The former phenomenon occures due to the formation of polynuclear complexes, such as $B_3O_3(OH)_4^-$, $B_4O_5(OH)_4^{2-}$, in concentrated boric acid solutions [9]. As for the latter fact the slope of the isotherm is defined as the distribution coefficient K_D,

$$K_D = [\overline{B}]/[B] \qquad (1)$$

where $[\overline{B}]$ and $[B]$ are the total concentration of boron in resin and in solution, respectively. The S-shape of the isotherm means that the distribution coefficient becomes small at either the low-concentration region or the high-concentration region.

In comparing Fig. 2 and 3, it is seen that the chromatograms in Fig. 2 are reflected with the nature of the ion exchange isotherms. The small distribution coefficient at low concentration leads the leakage like a nose in the adsorption stage and also the steep end of the rear boundary in the elution stage. On the contrary the large distribution coefficient at the higher but not very high concentration region results in the steep front boundary above ca. 0·2 M in adsorption stage and also the broadened boundaries at the corresponding region in the elution stage.

Isotope Separation Coefficient

Fig. 4—7 present the concentration of boric acid and the atomic fraction of ^{10}B isotope in the effluent at the rear boundary regions. The isotopic single stage separation

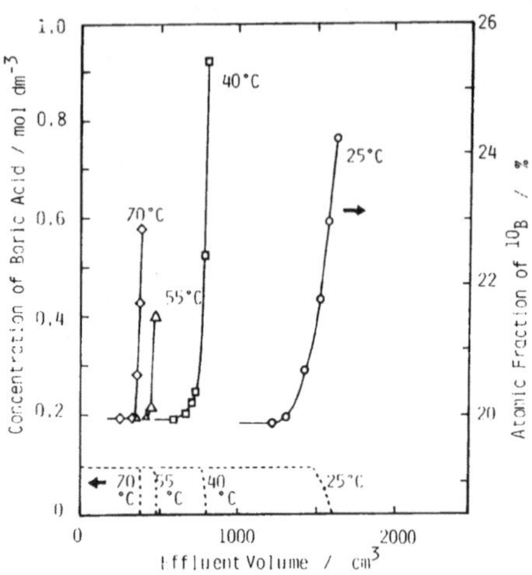

Fig. 4. Profiles of the eluted boric acid (dashed lines) and the ^{10}B enrichment at the rear boundary (solid lines)

Feed boric acid: 0·1 M

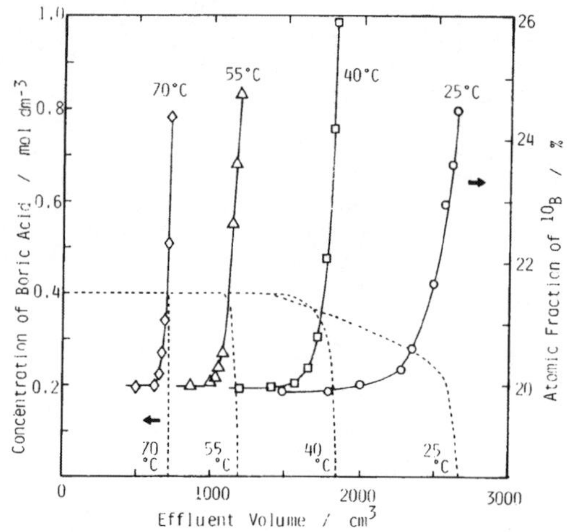

Fig. 5. Profiles of the eluted boric acid and the ^{10}B enrichment at the rear boundary

Feed boric acid: 0·4 M

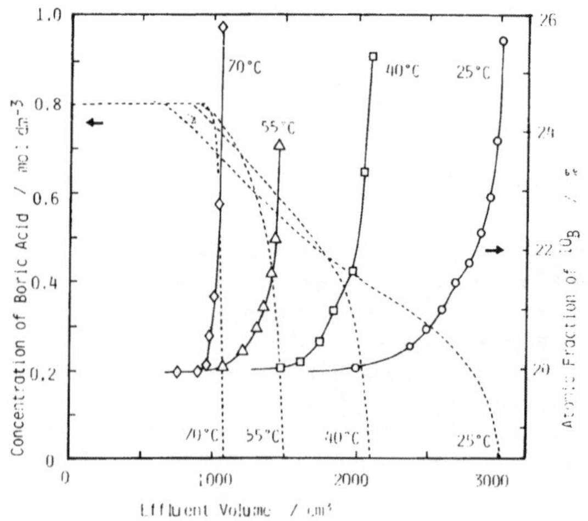

Fig. 6. Profiles of the eluted boric acid and the ^{10}B enrichment at the rear boundary

Feed boric acid: 0·8 M

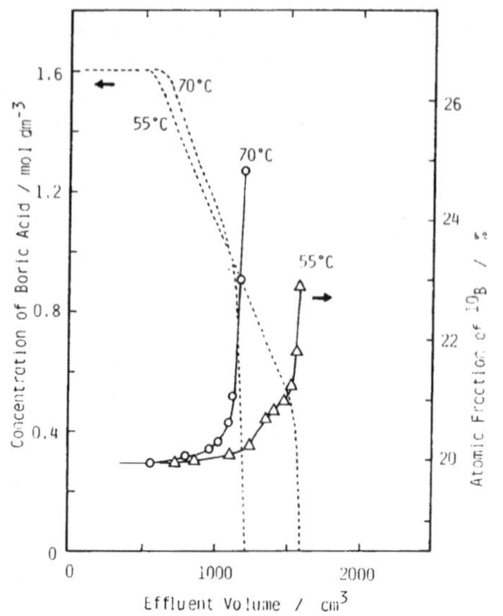

Fig. 7. Profiles of the eluted boric acid and the ^{10}B enrichment at the rear boundary

Feed boric acid: 1·6 M

factor is defined as

$$\alpha_s = [\overline{^{10}B}]\,[^{11}B]/[\overline{^{11}B}]\,[^{10}B] \qquad (2)$$

where $[\overline{B}]$ and $[B]$ are isotopic boron concentrations in resin and solution phases, respectively. The separation factor is experimentally determined by using the analysis data on the effluent.

$$\alpha_s = 1 + \varepsilon \qquad (3)$$

$$\varepsilon = \sum f_i(R_i - R_0)/QR_0(1 - R_0) \qquad (4)$$

where Q is the total ion exchange capacity for the isotopes, R_0 and R_i are atomic fractions of ^{10}B in the original feed and in the i-th sample in effluent, respectively, and f_i is the moles of total boron isotopes in the i-th sample. The deviation of the separation factor from unity is referred as isotope separation coefficient and expressed with ε. Using Eq. (4) the values of ε were calculated for all experimental systems and presented in Fig. 8.

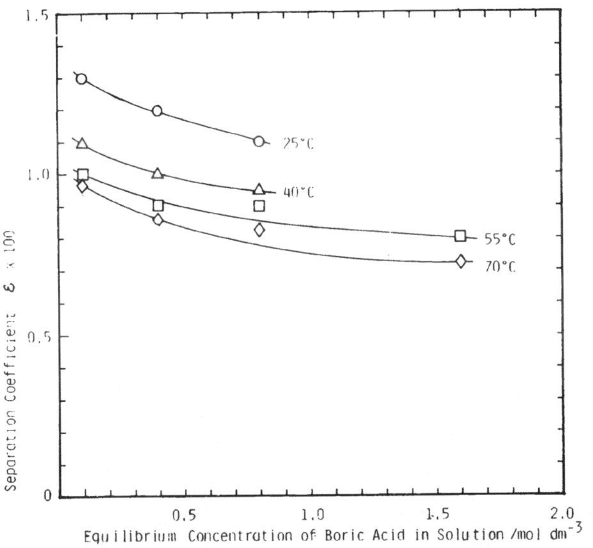

Fig. 8. Experimentally obtained isotope separation coefficients ε
ε = single stage isotope separation factor − 1

As seen in Fig. 8, the separation coefficient decreases with increases in temperature and also boric acid concentration. These tendencies are consistent with those determined by batchwise experiments [6] and with those obtained by using strongly basic anion exchange resins by *Rosset* [2] and *Schütze* [4]. In the present work the effect of the boric acid concentration is not so large as in the cases of the previous work above-mentioned. The rear boundaries were broadened specifically in the chromatograms at high concentrations of the feed boric acid. This fact suggests that ε's of these cases might be elevated from the expected values according to the lowered effective concentration of boric acid at the boundary region.

Optimum Condition for Enrichment

When the ion exchange chromatography is applied to the production of enriched isotopes, there are some requirements concerning the chromatographic properties:

1) large isotope separation coefficient of the system,
2) large ion exchange capacity for isotopic species,
3) steep isotope-accumulation curve in the band boundary region,
4) sharp-cut band boundary maintained by a self sharpening mechanism,
5) long life of ion exchange resin.

To consider the first two requirements listed above, we define here the specific enrichment capacity E

$$\left.\begin{aligned} E &= \Delta n/\tau V \\ \Delta n &= \sum f_i(R_i - R_0) = \varepsilon QR_0(1 - R_0) \end{aligned}\right\} \qquad (5$$

where Δn represents moles of enriched isotope (e.g. ^{10}B), τ is the effective time for production, here represented by the time for the elution process, and V is the volume of ion exchange resin used.

The calculated values of E for all experiments are plotted in Fig. 9. In spite of the relatively small differences in the separation coefficients shown in Fig. 8, the large variations were found in the magnitude of the specific enrichment capacities presented in Fig. 9. The results suggest the higher concentration of feed boric acid is a preferable condition for the production of ^{10}B-enriched boron.

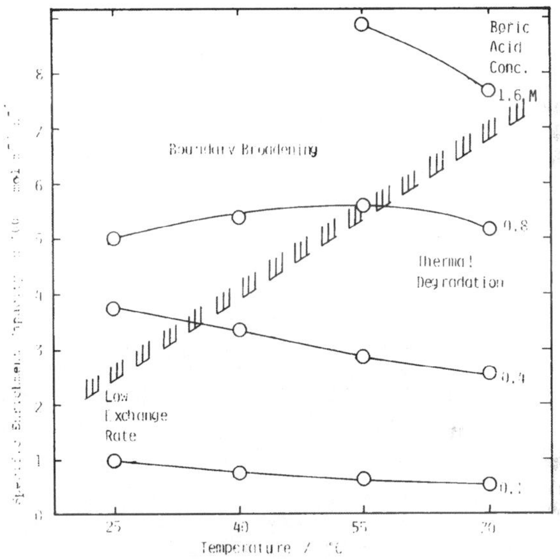

Fig. 9. Specific enrichment capacity of the system consisting of weakly basic anion exchange resin WA-21 and eluent of pure water
Undesiable conditons are indicated in the figure

The specific enrichment capacity determines the size of the plant conceived. The slope of the isotope-accumulation curve, on the other hand, determines mainly the duration of the equilibrium time, i.e. the time between the start-up of operation and the actual production of the enrichment. The larger slope of the accumulation curve can provide the better productivity saving the equilibrium time. Looking through Fig. 4—7, it is found that the experiments at 25 °C give less steepened isotope-accumulation curves. The higher temperatures than 25 °C are preferable for the production of enriched boron, accordingly.

A long distance displacement chromatography is indispensable for higher enrichment of isotopes, and is realized only when the boundary of the adsorption band is kept being sharp during the elution process. The sharp displacement boundary is produced by the one-side shifted ion exchange equilibrium for either adsorption or elution. These shifted equilibrium results in the self-sharpening mechanism at the band boundary. Fig. 4—7 indicate that the self-sharpening mechanism works well in the experiments at low concentrations of boric acid and at high temperatures where the distribution coefficients are small as previously discussed. The undesiarable conditions in respect of the band shape are marked in Fig. 9: the upper half part above the dashed slant line is the area of the broadened boundaries.

As for the fifth item in the listed requirements, the durability of anion exchange resins is very sensitive to the operating temperature. High temperature should be avoided for the operation. In the previous work of 256 m chromatography at 40 °C using the same type of resin, the ion exchange capacity decreased by about 20% after the chromatography. Considering these facts it is suggested to use this type resin at the lower temperature than 60 °C to avoid enormous degradation of the resin while in the operation.

In the light of the points discussed in this section it is concluded that the operational conditions of 40 °C and 0·4 M boric acid feed solution are the optimum for the production of enriched ^{10}B by ion exchange chromatography using weakly basic anion exchange resin, WA-21, within the operational conditions examined in the present work.

Received May 9, 1984

References

[1] *Y. Yoneda, T. Uchijima, S. Makishima*, J. Phys. Chem. **63** (1959) 2057

[2] *R. Rosset, H. Fould, M. Chemla* et al., Bull. Soc. Chim. Fr. **1964**, 607

[3] *M. M. Urgell, J. Iglesias, J. Casas* et al., Third UN Intern. Conf. on Peaceful Uses of Atomic Energy, May 1964, A/CONF. 28/P/491

[4] *G. Christoph, J. Heybey, H. Schütze* et al., Isotopenpraxis **12** (1976) 17

[5] *T. Hirao, M. Kotaka, H. Kakihana*, Nippon Kagaku Kaishi **1973**, 1477

[6] *H. Kakihana*, J. Chromatog. **102** (1974) 47

[7] *Y. Sakuma, M. Aida, M. Okamoto, H. Kakihana*, Bull. Chem. Soc. Jpn. **53** (1980) 1860

[8] *M. Nomura, M. Okamoto, H. Kakihana*, Shitsuryo Bunseki **21** (1973) 277

[9] *R. E. Mesmer, C. F. Baes Jr., F. H. Sweeton*, Inorg. Chem. **11** (1972) 573

Isotopenpraxis **21** (1985) 6, S. 208 bis 210

Pulse Radiolysis of Hydroquinone Solutions in the Presence of N_2O and O_2

A. A. Al-Suhybani

(Dept. of Chemistry, College of Science, King Saud University, Saudi Arabia)[1])

G. Hughes

(Dept. of Chemistry, Liverpool University, U.K)

The pulse radiolysis of hydroquinone solutions in the presence of a mixture of N_2O and O_2 has shown that a transient species is formed by the addition of O_2 to the trihydroxycyclohexadienyl radical to form a peroxy radical. The G value for this radical is estimated to be $\sim 6{\cdot}5$ and the rate coefficient of its decay is found to be $\sim 2{\cdot}7 \times 10^7\ M^{-1}\ s^{-1}$.

Die Impulsradiolyse von Hydrochion-Lösungen in Gegenwart eines Gemischs von N_2O und O_2 hat gezeigt, daß durch die Zuführung von O_2 zum Trihydroxycyklohexadienyl-Radikal eine Übergangsspezies gebildet wird und schließlich ein Peroxy-Radikal entsteht. Der G-Wert dieses Radikals wird auf ca. 6,5 geschätzt; es wurde ein Zerfallskoeffizient von ca. $2{,}7 \cdot 10^7\ M^{-1}\ s^{-1}$ ermittelt.

Keywords

electron beams; phenoxy radicals; pulse techniques; radiation chemistry; radicals; radiolysis

Introduction

The formation of trihydroxycyclohexadienyl radicals ($H_2\dot{Q}-OH$) by the addition of OH radicals to aromatic rings or of H atoms to hydroxybenzenes has been studied in a number of cases [1]. Thus the formation and decay of $H_2\dot{Q}-OH$ under different conditions has been reported [2, 3]. In the pulse radiolysis of hydroquinone (H_2Q), it has been shown [3] that the rate of decay of $H_2\dot{Q}-OH$ depends on $[O_2]$ at low $[H^+]$ being faster at higher $[O_2]$. This was explained as being due to the reaction of $H_2\dot{Q}-OH$ with O_2 and the rate coefficient for this reaction was determined to be $1{\cdot}21 \times 10^9\ M^{-1}\ s^{-1}$. This value compares with $6{\cdot}1 \times 10^8\ M^{-1}\ s^{-1}$ reported by others [4]. In this work, pulse radiolysis was used to study the kinetic behaviour of the peroxy radical formed as a result of the addition of OH radicals to H_2Q and the subsequent reaction with O_2 in a 1 : 1 mixture of N_2O and O_2.

Experimental

The present work was carried out at the Patterson Laboratories, Christie Hospital and Holt Radium Institute, Manchester. A Vickers 12 MeV Linear Accelator was used to produce short pulses of electrons from 5 ns to 5 µs. The transient species were detected by optical methods. The amount of absorbed energy was determined by the oxidation of 10^{-3} M ferrocyanide in neutral, O_2, saturated solutions. The ferricyanide produced was measured spectrophotometrically at 420 nm where its extinction coefficient was taken to be $10^3\ M^{-1}\ cm^{-1}$ [5]. In calculating dose, $G_{OH} = G$ (ferri.) $= 2{\cdot}75$ in neutral solutions was assumed. As an example, the dose per 20 ns pulse $= 7{\cdot}45 \times 10^{19}$ eV/L. The reproducibility was good provided that all experimental conditions were kept the same. The ge-

[1]) Address: Riyadh-11451, Saudi Arabia (all correspondence to this address)

neral pulse radiolysis technique and methods of handling solutions have been described elsewhere [6].

Materials

In all experiments, triply distilled water was used. H_2SO_4 was 95–98% BDH and hydroquinone was an Eastman product. Both were analar grade. Argon and oxygen were from B.O.C. and N_2O was supplied by Matheson or Air Products. All these gases were washed before use.

Results and Discussion

Fig. 1 shows the spectra produced in 5×10^{-4} M H_2Q/10^{-4} M HCl, Ar saturated solution following a 20 ns pulse, taken at longer time scales (ms). Similar spectra are produced when the solution is saturated with N_2O as can be seen in Fig. 2. It is clear that D_{max} in N_2O solutions is ~5–8 times higher than D_{max} in Air saturated solutions. In addition the decay of the spectrum becomes slower in N_2O saturated solutions. It was, in fact too slow to be measured.

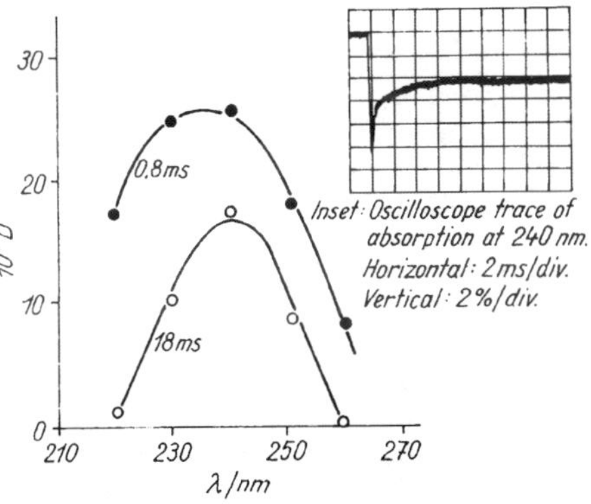

Fig. 1. Absorption spectra produced in 5×10^{-4} M H_2Q/10^{-4} M HCl, Ar saturated
Pulse length 20 ns

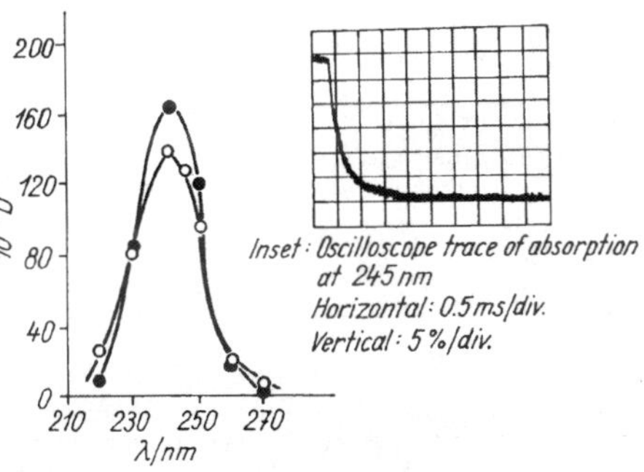

Fig. 2. Absorption spectra produced in 5×10^{-4} M H_2Q/10^{-4} M HCl, N_2O saturated
Pulse length = 20 ns; ○ – after 0·8 ms; ● – after 18 ms

The formation of the spectrum is by a first order process with $k = 3{\cdot}5 \times 10^3\ s^{-1}$ as can be seen in Fig. 3. This value is one order of magnitude less than the rate for the uncatalysed decomposition of trihydroxycyclohexadienyl radical, i.e. the reaction

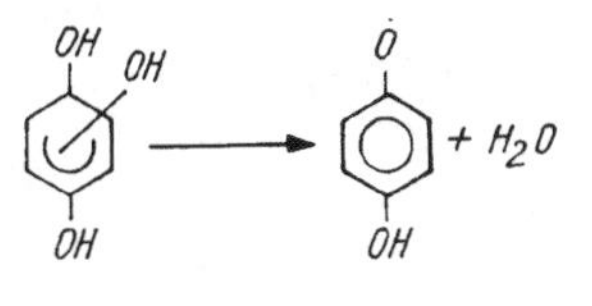 (1)

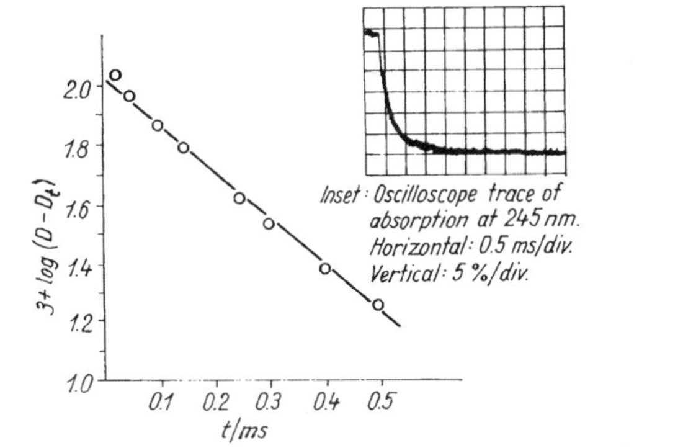

Fig. 3. 1st Order plot for the formation of the absorption in 5×10^{-4} M H_2Q/10^{-4} M HCl, N_2O saturated
Pulse length = 20 ns

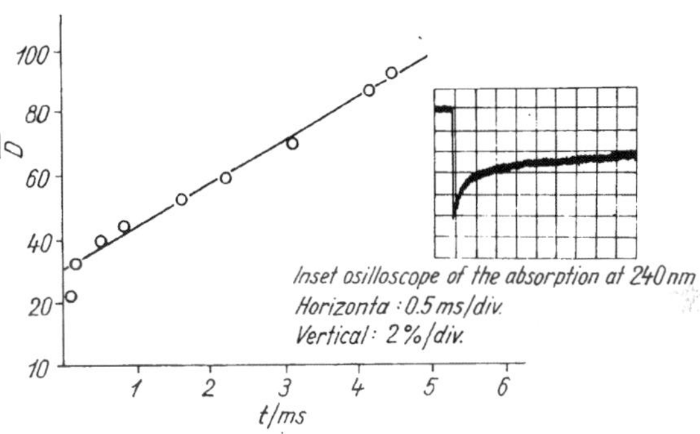

Fig. 4. 2nd Order plot for the decay of the absorption formed in 5×10^{-4} M H_2Q/10^{-4} M HCl
Ar saturated

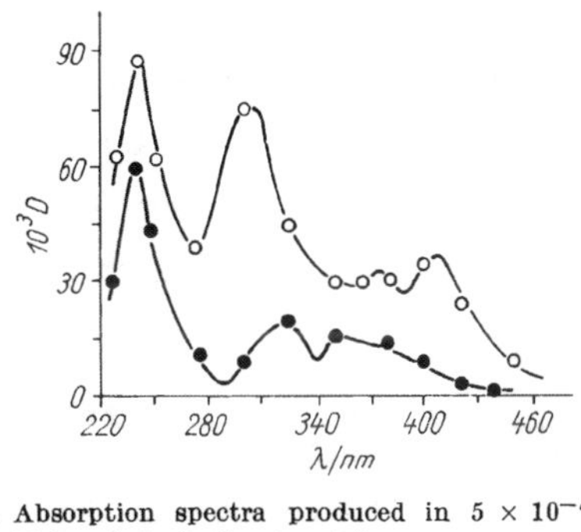

Fig. 5. Absorption spectra produced in 5×10^{-4} M H_2Q/10^{-4} M HCl, N_2O and O_2 (1:1) saturated
Pulse length = 20 ns; ○ – 20 μs; ● – 1 ms

In Ar saturated solution and at ~240 nm the decay is by second order kinetics as can be seen in Fig. 4, but in N_2O saturated solutions the spectrum becomes relatively stable (~18 ms).

If, however, the solution is saturated with a mixture of N_2O and O_2 (1:1) we obtain the spectra showsn in Fig. 5 taken 20 μs and 1 ms after a 20 ns pulse. The part of the spectrum around 400 nm is similar to that attributed to $H\dot{Q}$ in Ar or O_2 saturated solution. The part around 240 nm is similar to that produced in Ar or N_2O and shown in Fig. 1 and 2. The peak at 300 nm could not be attributed to a particular species, since $H_2\dot{Q}-OH$ has its peak at $\lambda < 320$ nm [2] and $H\dot{Q}$ (or $\dot{\bar{Q}}$) is also reported to have strong absorptions in this region of the spectrum [7].

The following analysis shows that the spectrum produced in N_2O and O_2 mixture is not due to a single species. At 420 nm 70% of the absorption decays by second order kinetics with a slope of $6 \cdot 1 \times 10^5\ s^{-1}$ while the remaining 30% decays with a slope of $3 \cdot 1 \times 10^5\ s^{-1}$. Lowering the dose rate to half (10 ns) does not affect the observed decay rate. At 300 nm 70% decays in $\sim 140\ \mu s$ with a slope of $2 \cdot 4 \times 10^5\ s^{-1}$, the remaining part decays in much longer time (1·7 ms) and the kinetics decay is neither first not second order. At 240 nm $\sim 40\%$ of the absorption decays by second order in $\sim 1 \cdot 7$ ms (limit of the measurement), the slope being $5 \cdot 4 \times 10^3\ s^{-1}$.

Since quinone (Q) has its peak at ~ 245 nm [8] one might expect that the spectrum at ~ 240 nm to be due to Q but this can be ruled out for the following reasons:

1) The formation of the spectrum is first order while the formation of Q is second order, i.e. through the reaction:

$$H\dot{Q} + H\dot{Q} \rightarrow Q + H_2Q \qquad (2)$$

2) It is found that $G(Q)_{observed}$ is much less than $G(Q)_{calc.}$ in both Ar and N_2O saturated solution, if reaction (2) is the only source of Q.
3) If the spectrum is due to Q, then the most likely species to react with it are e_{aq}^- and H and the measured rate coefficient for its disappearance is much less than the reported values for the reactions of these species with Q [9].
4) At 400 nm D_{max} in N_2O solutions is approximately double that in Ar solutions. This would be expected since the following reaction is occurring:

$$e_{aq}^- + N_2O + H_2O \rightarrow OH + OH^- + N_2 \qquad (3)$$

On the other hand D_{max} at 240 nm is > 5 times higher in N_2O solution.
5) If $H\dot{Q}$ is the precursor for Q, then its decay should be the same in Ar, N_2O and a mixture of N_2O and O_2. This is not the case.

It is possible that in Ar and N_2O saturated and high pH an alternative reaction of $H_2\dot{Q}-OH$ takes place

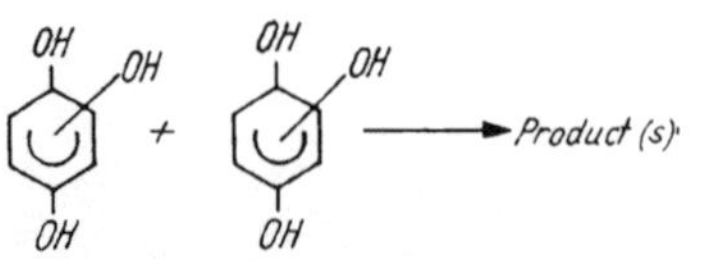

(4)

The difference in the decay rate between Ar and N_2O might be attributed to the reaction of e_{aq}^-, in case of Ar solutions, with some of the intermediates.

In the presence of a mixture of N_2O and O_2 $\sim 90\%$ of e_{aq}^- will react with N_2O ($[N_2O] = 10^{-2}$ M) according to reaction (3). If $G_{e_{eq}^-} = 2 \cdot 8$, then $G \sim 5 \cdot 5$ of OH will be available to react with H_2Q according to the reaction [2]

$$H_2Q + OH \rightarrow H_2\dot{Q}-OH \qquad (5)$$

The remaining e_{aq}^- and H will react with O_2 to form $H\dot{O}_2$ ($\bar{O}_2$). Under the present conditions and taking k_{obs} for the water elimination reaction as $4 \cdot 7 \times 10^4\ s^{-1}$, then $t_{1/2}$ of reaction (1) will be $\sim 15\ \mu s$ while $t_{1/2}$ for reaction of $H_2\dot{Q}-OH$ with O_2 will be $\sim 1\ \mu s$ [3]. Thus under the conditions of this experiment, virtually all OH adduct reacts with O_2 and virtually none forms $H\dot{Q}$ by water elimination. Therefore, it is quite reasonable to assume that the observed spectrum is predominantly that of the peroxy radical. It is very likely that the residual absorption at 240 nm is due tothis peroxy radial, since it has been reported that most organic peroxy radicals have absorptions in this region of the spectrum with extinction coefficients between $1-2 \times 10^3\ M^{-1}\ cm^{-1}$ [10]. For example, the radiacals $O_2C_6H_{11}OH$ and $O_2C_6H_{11}$ have absorption maxima at 246 and 255 nm and extinction coefficients of 1·6 and $1 \cdot 9 \times 10^3\ M^{-1}\ cm^{-1}$ respectively [11]. By analogy with other peroxy radicals, the radical produced in the reaction,

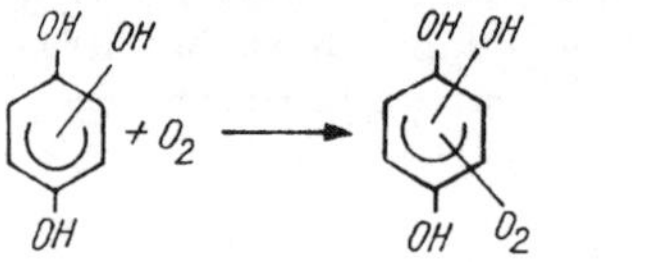

(6)

will decay by second order kinetics, e.g.

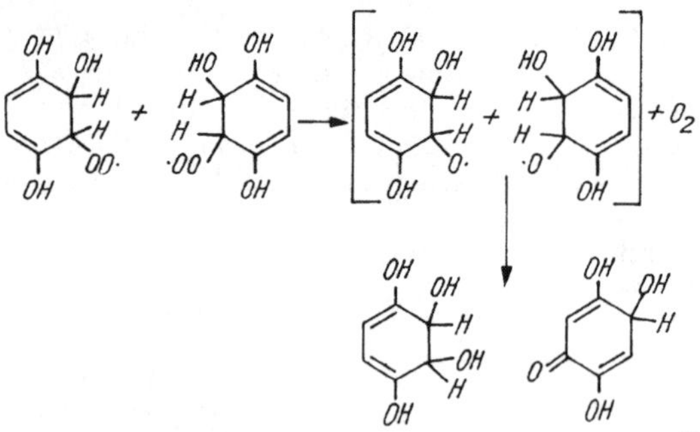

(7)

The products of this reaction may be unstable and undergo further decomposition. If it is assumed that only reaction (6) occurs and assuming the radical to be a peroxy radical, then by taking ε_{max} to be 2000 $M^{-1}\ cm^{-1}$, the rate coefficient of the bimolecular decay, reaction (7) will be $\sim 2 \cdot 7 \times 10^7\ M^{-1}\ s^{-1}$ which is not very different from that of, say, $O_{26}CH_{11}OH$ ($2k = 1 \cdot 2 \times 10^7\ M^{-1}\ s^{-1}$) taking into account the uncertainty in ε and the effect of solvent. Again if only reaction (6) is considered, then a rough estimate of $G[H_2Q-(O_2)OH]$ can be obtained. From the known dose and assuming ε_{max} $\sim 2000\ M^{-1}\ cm^{-1}$ (at 240 nm), $G[H_2Q(O_2)OH] \sim 6 \cdot 5$ was found. This is quite reasonable, if comparedi with $G(OH)_{calc.} = 5 \cdot 5$.

Acknowledgement

We are grateful to the late Professor *J. H. Baxendale* for help with the experimental work.

Received March 28, 1984

References

[1] *E. J. Fendlier, J. H. Fendlier*, Prog. Phys. Org. Chem. **7** (1970) 267
[2] *G. E. Adams, B. D. Michael*, Trans. Faraday Soc. **63** (1967) 1171
[3] *A. A. Al-Suhybani, G. Hughes* (to be published)
[4] *O. J. Micic, M. T. Nenadovic*, J. Phys. Chem. **80** (1976) 940
[5] *M. S. Matheson, L. M. Dorfman*, Pulse Radiolysis, The M.I.T. Press (1969) p. 64
[6] *J. P. Keene*, In Pulse Radiolysis, Edited by *M. Ebert, J. P. Keene, A. J. Swallow* and *J. H. Baxendale*, Academic Press, London 1965, p. 1
[7] *S. I. Fukuzumi, Y. Ono, T. Keii*, Bull. Chem. Soc. Jap. **46** (1973) 3353
[8] *R. D. Bauman*, Absorption Spectroscopy, John Wiley & Sons, New York 1962
[9] *R. L. Willson*, Trans. Faraday Soc. **67** (1971) 3020
[10] *M. Simic*, Fast Processes in Radiation Chemistry and Biology. *G. E. Adams, E. M. Fielden* and *B. D. Michael*, Eds., Institute of Physics, London and Wiley-Interscience, London and New York 1975, p. 162
[11] *M. Simic, E. Hayon*, J. Phys. Chem. **75** (1971) 1677

Mitteilungen / Information

IAEA: VERBESSERUNG DER SCHAF- UND ZIEGENZUCHT

In Asien, dem Nahen Osten, Afrika und Lateinamerika gibt es etwa 400 Millionen-Schafe und mehr als 300 Millionen Ziegen. Sie sind wichtige Lieferanten fuer Fleisch, Milch, Wolle und Produkte, die fuer den einheimischen Verbrauch nuetzlich sind und im Prinzip keine Infrastruktur, Maerkte oder externe Technologien erfordern. Im Jahre 1983 hat die>IAEA zusammen mit der FAO aus diesem Grund ein koordiniertes Forschungsprogramm ueber problemorientierte Forschung zur Verbesserung der Schaf- und Ziegenzucht ins Leben gerufen. Bei einer juengst stattgefundenen Forschungskoordinationstagung in Nairobi, Kenia vom 4.-8.3. haben Wissenschaftler aus der VR China ueber die Fortpflanzungsfaehigkeit der gekreuzten Saanen Chengdu Mah Ziege berichtet. Forscher aus Malaysia haben ueber Versuche informiert, die Zunahme des Lebendgewichts der Katjang Ziege zu verbessern. Dies soll mit Hilfe einer Reisstrohfuetterung geschehen, die mit Saeure oder Alkali angereichert wurde, um den Naehrwert zu erhoehen. Ziegen und Schafe werden oft in Gebieten gezuechtet, in denen ein rauhes Klima mit extremer Hitze und/oder Kaelte und hohe Feuchtigkeit herrscht, und auf mageren Weiden mit niedrigem Naehrwert, die oft von Krankheitserregern stark verseucht sind. Auch Wassermangel tritt periodisch auf. Koennten die Zuchtergebnisse der kleinen Wiederkaeuer innerhalb kleiner Farmbetriebe gesteigert werden, dann wuerde die Ernaehrung von Millionen von Menschen in einigen der aermsten Gebiete der Welt um vieles verbessert werden. Die Steigerung der Zuchtergebnisse in diesen Gebieten erfordert bessere Ernaehrung der Schafe und Ziegen, erhoehte Fortpflanzung und eine verstaerkte Krankheitsbekaempfung. Es gibt wertvolle Verfahren von Kerntechniken fuer jedes dieser Gebiete. Isotopische Methoden koennen angewendet werden, um den Naehrwert des an Ort und Stelle vorhandenen Futters zu ueberpruefen. Radioaktive Isotopenindikator-Technologie dient zur exakten Bestimmung der Wasserbeduerfnisse der Schafe und Ziegen, die unter verschiedenen Umweltbedingungen gezuechtet werden, und um wirkungsvolle Bewaesserungsmethoden ins Auge zu fassen. Analytische Methoden, die auf Isotopenindikatoren beruhen (z.B. Radioimmunoassay) haben einen bes. Wert bei Tierfortpflanzungsstudien. Diese Techniken sind einfach, billig und schnell, da die Messungen an Milch- oder Blutproben durchgefuehrt werden. (IAEA-Pressekommunique (1985) 4)

Kurznachrichten / Newlets

Radioactivity:

ALIMARIUS COEFFICIENT - A UNIVERSAL UNIT OF SPECIFIC ACTIVITY

The growing importance of specific activity concerns especially surface phenomena, toxicity, labelling of radiopharmaceuticals, isotope exchange, enzymatic and pharmacological ligand-receptor reactions. Analyzing the present state of evaluating the specific activity the author another time proposes the introduction of the D_{CF}-coefficient (deviation from the tone carrier free state) as a posibility for universal declaration of the specific activity and to overcome the great diversity in this field. The coefficient is defined as the negative decadic logarithm of the specific activity ratio $D_{CF} = - \lg N^x/N$; N^x- number of radioactive atoms, N-total number of atoms of the relevant element. For a quick estimation nomograms of the number of g-atoms and the weight of an element corresponding to the unit of radioactivity of the pure radionuclides in dependence on half-life and atomic mass can be reproduced. The dimensionless unit ranges between reasonable values (something between 0 and 100). D_{CF} migth be a solution for a simple, clear and easy-to-handle value defining the specific activity of the distinct preparation destinated for any purpose. Should D_{CF} come into practice, the author proposes to call it "Alimarius coefficient" in order to honour his great contribution to development of radioanalytical chemistry. (K. Svoboda, Radioanalytical and Nuclear Chemistry, Articles 88 (1985) 1, S. 13)

Trace analysis:

SUBSTOICHIOMETRIC ANALYSIS IN THE USSR

The principle of substoichiometry found a wide application in isotope dilution and activation analysis eliminating the necessity of the determination of the specific activity of the compounds separated or of the chemical yields, respectively. In the Soviet Union, the studies in the field of substoichiometric analysis were directed toward the development of new variants to improve the sensivity, reproducibility and selectivity of radioanalytical methods, the search of highly sensitive reagents for substoichiometric separation of trace elements, the choice of effective two-phase systems ensuring quantitative separation of two forms of the element to be determined and the investigation of possibilities for simultaneous determinatio of several elements. Abount 250 papers on the development of the theory of substoichiometry and its practical application have been published up to now. The application has been intensively increased especially in the period since the 1970's to date. The period of doubling is 2.5 years. The largest portion of works deals with application of the substoichiometric principle in the field of radiotracer methods (85%), and only 15% of the papers concern substoichiometry in neutron activation analysis. Summing up, it is obvious that substoichio metry is a method under development. (G.N. Bilimovich, J. Radioanalytical and Nuclear Chemistry, Articles 88 (1985) 1, S. 171)

Nuclear Decay:

AN EXOTIC RADIOACTIVE DECAY OF RA-223

A very interesting experiment has been reported recently. By using a magnetic superconducting spectrometer and an intense source of Ac-227, Gales et al. [Phys. Rev. Letters 53(1984)259] succeeded to obtain, in a run of 5 days, eleven events confirming the discovery of C-14 radioactivity of Ra-223, which was published in 1984 by Rose and Jonas [Nature 307(1984)245]. This is the second confirmation, the first being done in Moscow [D.V. Aleksandrov et al., Pis'm v JETF 40(1984)152] in a run of 30 days. The branching ratios relative to α-decay determined by these groups are in agreement with the value reported by Rose and Jones: $(8,5 \pm 2,5) . 10^{-10}$ from which one gets $\sim 10^{15}$s for the Ra-223 half life against thir natural radioactivity process. (W. Greiner et al., Z. Physik A320 (1985) 347-48)

Radiation exposures:

DOSE LIMIT CONDITIONS REVISED

The National Radiological Protection Board (NRPB) has strengthened its previous advice and now recommends that radiation exposure of the public should be controlled deliberately to ensure that radiation doses to individual members of the public do not exceed 70 mSv (7rem) in a lifetime or 1 mSv (0,1 rem) per year on average. The International Commission on Radiological Protection (ICRP) issued a statement, after its meeting in Washington in October 1983, relating to the limitation of doses to individual organs or tissues of members of the public in the context of expected lifetime exposures. ICRP has defined two types of biological effects resulting from exposure to ionising radiation. Stochastic effects are those for which the probability of an effect occurring, rather than its severity, is regarded as a function of dose, without threshold. Nonstochastic effects are those for which the severity of the effect varies with the dose, and for which there is a threshold of dose below which no effect is observed. In ICRP's opinion, the limitation of lifetime effective dose equivalents to members of the public to 70 mSv is sufficient to provide compliance with the limit for single organs, thus avoiding non-stochastic effects. The NRPB therefore agrees with ICRP and recommends that, apart from a few special cases, the limit on single organ doses is no longer necessary to prevent non-stochastic effects. NRPB-ASP6. The non-stochastic dose limit for members of the public, ISBN. (Atom 339 (1985) 1, 35 p.)

Prizewinner:
FRENCHMAN AWARDED ENRICO FERMI PRIZE
Georges Vendryes has become the first person outside the USA to be awarded the annual Enrico Fermi prize since its increption in 1954. The award has been made in recognition of his contribution to the French nuclear energy programme. Mr Vendryes is Scientific Director at the Commissariat à l'Energie Atomique. (Atom 339 (1985) 1, p. 27)

Kernanlagen:
SICHERHEITSKONTROLLE IN DER SOWJETUNION
Nach Ankuendigung durch den Sowjetischen Aussenminister Andrei Gromyko auf der 2. Sondertagung der Generalversammlung der Vereinten Nationen ueber Abruestung im Juli 1982 und den darauffolgenden Verhandlungen zwischen der Sowjetunion und der IAEA, haben die Sowjetunion und die IAEA am 21.2.1985 ein Sicherungs-Kontrollabkommen fuer die von der Sowjetunion bestimmten friedlichen Kernanlagen unterzeichnet. Seitens der IAEA wurde das Abkommen von Generaldirektor Dr. Hans Blix und seitens der Sowjetunion von Herrn A.M. Petrosyants, Vorsitzender des staatlichen Ausschusses ueber Anwendung von Kernenergie der UdSSR, unterzeichnet. Nach Inkrafttreten des Abkommens wird die IAEA ueber das gesamte Kernmaterial, das sich in Anlagen befindet, welche die IAEA von einer Liste der Anlagen, die die Sowjetunion fuer diesen Zweck bestimmt hat, auswaehlen wird, Sicherheitskontrolle ausueben. (IAEA-Pressekommunique (1985) 3)

Neuartige Strahlentherapie:
BOR UND NEUTRONENSTRAHLEN
Der Einsatz von B und Neutronenstrahlen ist eine der wenigen Behandlungsmoeglichkeiten fuer die Menschen, die an einem Gehirntumor leiden. Die dem Patienten verabreichte B-Verbindung wird von den Krebszellen im Gehirn angezogen. Durch anschliessende Neutronenbestrahlung wird das B radioaktiv. Das Problem bei dieser Art der Behandlung besteht darin, die maximale Dosis auf die Krebszellen zu konzentrieren, ohne das gesunde Gewebe zu beschaedigen. Die Ende der fuenfziger Jahre in Grossbritannien erprobte Bortherapie wurde aufgrund der Schwierigkeiten, die Dosis zu steuern, wieder abgebrochen. Zwei Wissenschaftler des Kernforschungslabors der Zentralen Elektrizitaetsbehoerde Grossbritanniens in Berkeley haben jetzt eine Methode entwickelt, bei der energieschwache Neutronen in einer Folge von kurzen Stoessen zur Kopfmitte dringen. Das Forschungslabor der Atomenergiebehoerde (UKAEA) in Harwell hat die einzigen Reaktoren, die eine geeignete Dosis produzieren koennen. Daher arbeiten zwei grosse Krankenhausgruppen in der nahegelegenen Stadt Oxfort in diesem Bereich der Forschung mit Harwell und den Labors von Berkeley zusammen. (Atomwirtschaft/Atomtechnik 30 (1985) 2, S. K+U III)

Energiewirtschaft:
STEIGENDER ENERGIEVERBRAUCH
Nach einer Uebersicht der ESSO AG hat sich der Primaerenenergieverbrauch in der Welt 1983 zum ersten Mal seit 4 Jahren wieder erhoeht. Er erreicht ca. 8,635 Mrd. t Steinkohleneinheiten (SKE) und war damit immer noch unter dem bisherigen Hoechststand von 8,850 Mrd. SKE, die 1979 verbraucht wurden. Durch sparsameren Umgang mit Energie und bessere Energienutzung vermindert sich der Pro-Kopf-Verbrauch. So verbrauchte 1983 jeder Einwohner der BRD mit knapp 6 t SKE 6% weniger als 1980. Da die Bevoelkerungsstatistik mit einem Anstieg der Weltbevoelkerung auf etwa 7 Mrd. Menschen bis zum Jahre 2000 rechnen, ist in den kommenden Jahren eine weitere Erhoehung des Energiebedarfs in der Welt zu erwarten. Bei unveraendertem Energieverbrauch pro Kopf wuerde der Primaerenergieverbrauch bis zum Jahre 2000 um ca. 50% auf knapp 13 Mrd. t SKE steigen. Da jedoch bei weiterem technischen Fortschritt Energie immer rationeller verwendet werden kann, rechnet man bis dahin nur mit einem Mehrbedarf von 25 bis 30%, immerhin 2 bis 2,5 Mrd. t SKE. (Atomwirtschaft/Atomtechnik 30 (1985) 2, S. K+U III)

Kernenergie:
NIEDERLANDE FUER KERNENERGIE-AUSBAU
Nach mehrjaehriger breiter gesellschaftspolitischer Diskussion hat sich die niederlaendische Regierung unter Premierminister R. Lubbers Mitte Januar 1985 fuer den Ausbau der Kernenergie ausgesprochen. In den Niederlanden sind ein kleines Kernkraftwerk mit einer Leistung von 55 MWe und einem Siedewasserreaktor sowie ein weiteres mit 477 MWe und einem Druckwasserreaktor seit 1968 bzw. seit 1973 in Betrieb. Im Rahmen des neuen Energieprogramms bis 1995 sollen zwei weitere Kernkraftwerke mit einer Leistung von 900 bis 1300 MWe je Block errichtet werden. Elektrizitaetswirtschaft und Industrieunternehmen versprechen sich davon Wirtschaftlichkeit der Stromerzeugung und Beschaeftigungseffekte. (Atomwirtschaft/Atomtechnik 30 (1985) 3, S. K+U III)

FRANKREICH-60% ATOMSTROM
Der Anteil des Kernenergiestroms an der gesamten franzoesischen Stromerzeugung lag im Jahre 1984 bei 59%. Die staatliche Elektrizitaetsgesellschaft Electricite de France (EdF) teilte mit, dass 1984 etwa 60% der Eigenerzeugung ihrer Kraftwerke Kernenergiestrom war. Mit einer Gesamterzeugung von erstmals ueber 200 000 Mrd. kWh (BRD 1984: 92,6 Mrd. kWh Kernenergiestrom) lagen die franzoesischen Kernkraftwerke hinter den USA wieder an zweiter Stelle der Welt. Nach ersten Statistiken erzeugten die Kernkraftwerke in den USA 1984 rund 335 Mrd. kWh. (Atomwirtschaft/Atomtechnik 30 (1985) 3, S. K+U IV)

ENERGIEKONZEPTION BIS ZUM JAHRE 2000 BERATEN
In der Zeit vom 14.-16. Januar 1985 fand in Moskau die 113. Sitzung des Exekutivkometees des RGW statt. Das Exekutivkomitee behandelte eine Reihe von Fragen der Realisierung der Beschluesse der Wirtschaftsberatung der Mitgliedslaender des RGW auf hoechster Ebene und der XVIII. (ausserordentlichen) Ratstagung.
In Verwirklichung der abgestimmten Strategie der Zusammenarbeit der RGW-Laender in den Energie-, Brenn- und Rohstoffbereichen billigte das Exekutivkomitee die Konzeption fuer die perspektivische Entwicklung der Elektroenergiewirtschaft im Rahmen der Vereinigten Energiesysteme der RGW-Laender fuer den Zeitraum bis zum Jahre 2000. Die Struktur der Erzeugung von Elektro- und Waermeenergie soll auf der Grundlage der vorrangigen Entwicklung der Kernenergetik, einer besseren Nutzung der eigenen Ressourcen jedes Landes, darunter kalorienarmer Brennstoff und neuer nichtkonventioneller Energietraeger, sowie der Reduzierung des Verbrauchs von fluessigem Brennstoff in Kraftwerken weiter vervollkommnet werden. Gebilligt wurden Massnahmen zur Erweiterung und Verbesserung des gegenseitigen Austausches von wissenschaftlich-technischen Informationen zwischen den RGW-Laendern. (Kommunique ueber die 113. Sitzung des RGW-Exekutivkomitees aus Leipziger

Volkszeitung (1985) 18. Januar)

Strahlentechnik:
JETZT Cs-137 FUER KLAERSCHLAMMBESTRAHLUNG
Die seit 10 Jahren in Betrieb befindliche Klaerschlammbestrahlungsanlage in Geiselbullach, in der Naehe von Muenchen, die bisher mit Co-60-Strahlenquellen arbeitet, wird seit 1984 schrittweise auf Cs-137-Strahlenquellen umgestellt. Dieses Radionuklid hat gegenueber Co-60 (HWZ 5,6a) eine wesentlich laengere Halbwertszeit (13,17a), allerdings eine geringere Gammaenergie (0,68 MeV) gegenueber Co-60 (1,1 MeV). Zunaechst wurden neun in Edelstahl eingeschweisste Cs-137-Quellen mit einer Aktivitaet von 63 000 Ci eingebaut. Die Gesamtenergie nach vollstaendiger Umruestung soll etwa 2 Mio. Ci betragen. Der betriebswirtschaftliche Vorteil bei voelliger Umruestung auf Cs-137-Strahlenquellen soll jaehrlich 200 000 DM betragen. (Atomwirtschaft (1985) 3, S. 119)

Strahlenbelastung:
NCRP-BERICHT UEBER RADON IN DEN USA
Der National Council on Radiation Protection and Measurement (NCRP) hat einen Bericht "Evaluation of Occupational and Environmental Exposures to Radon and Radon Daughters in the United States" veroeffentlicht. Der Bericht behandelt die Radon-Quellen in der Atmosphaere, die Dosimetrie fuer Inhalationen in Uranbergwerken, die Dosimetrie von Inhalationen von Radon-Zerfallsprodukten in der Umgebungsatmosphaere, das Ungleichgewicht von Radon-Zerfallsprodukten in der Atmosphaere, die Festlegung von Grenzbelastungen am Arbeitsplatz, Moeglichkeiten der Induktion von Lungenkrebs durch Radon-Zerfallsprodukte, Tierversuche zum Studium der Inhalationseffekte durch Radon-Zerfallsprodukte sowie das Risiko von Beschaeftigten in der Uranindustrie und der Allgemeinbevoelkerung durch Radon und seine Zerfallsprodukte. (Atomwirtschaft (1985) 3, S. 119)

Landwirtschaft:
VIELSEITIGE ANWENDUNG DER STRAHLENTECHNIK
Sowjetische Wissenschaftler von bedeutenden, im Artikel genannten wiss. Forschungseinrichtungen haben die folgenden Anwendungsmoeglichkeiten erfolgreich untersucht, die bereits zum groessten Teil in Pilotanlagen angewendet werde.
Bestrahlung von Setzlingen im Weinbau mit solchen Strahlendosen, die die Gewebeunvertraeglichkeit beim Pfropfen abschwaechen und das Verwachsen von Unterlagen und Pfropfreis foerdern. Die Arbeitsproduktivitaet stieg auf das 25fache, die Moeglichkeit des Befalls mit Bakterienkrebs verringert sich, und die Ertraege stiegen.
Durch Strahlenstimulierung von Eiern vor dem Brueten koennen jaehrlich 0,2-0,8% Kueken mehr produziert werden.
Kornkaefer und andere Schaedlinge werden mittels ionisierender Strahlung vollstaendig unterdrueckt. Die Selbstkosten betragen etwa die Haelfte gegenueber der Behandlung mit Methylbromid.
Durch Strahlensterilisierung von Insektenmaennchen koennen z.B. der Bohnensamenkaefer, die Kohleute und der Apfelwickler zuverlaessig bekaempft werden.
Die Bestrahlung von verschiedenen Obst- und Gemuesesorten sowie Lebensmitteln hat zur Verlaengerung der Lagerfristen gefuehrt. Es konnten weder schaedliche Effekte noch Reststrahlung nachgewiesen werden.
Seit mehreren Jahren wird am Institut fuer Biophysik der AdW der UdSSR die Wirkung von kleinen Dosen ionisierender Strahlung erforscht. Es wurde festgestellt, dass sie das Erscheinen des genetischen Programms im Organismus stimulieren.
So erhoehen sich die Ertraege von bestrahlen Saat- bzw. Pflanzgut z.B. bei Getreide und Koernerleguminosen um 10-15%, bei Gemuese um 20-30%,und bei Kartoffeln um 15-20%; so enthalten aus bestrahltem Saatgut gezogene Moehren mehr Karotin und Zuckerrueben mehr Saccharose.
(A. Kusin u.a., Presse der SU (1985 1, S. 35, nach Prawda v. 23.10.1984)

Veröffentlichungen/
Publications

N e u e r s c h e i n u n g :

Proc. XXII Internat. Conference on High Energy Physics, Leipzig, July 19-25, 1984
(Editors: A. Meyer and E. Wieczorek, Institut fuer Hochenergiephysik, DDR-1615 Zeuthen, Volumen I, 381 pp.; II, 376 pp.)

Akademie-Verlag, Berlin
Niels Bohr - Leben und Werk einer Atomphysikers
(U. Roeseberg, 1985, ca. 320 S.)
Hopping Conduction in Solids
(H. Boettger and V.V. Bryksin, 1985, ca. 380 pp.)
Quantum Statistics of Charged Particle Systems
(W.-D. Kraefft, D. Kremp, W. Ebeling and G. Roepke, 1985, ca. 320 pp.)
Struktur von Festkoerpern der Mikroelektronik Werkstoffe von LSI-Schaltkreisen
(F.L. Edelmann, 1985, ca. 240 S.)
Rauschdiagnostik
(P. Liewers, 1985, 234 S.)
Theorie chemischer Elementarprozesse
(E.E. Nikitin u. L. Zülicke, 1985, ca. 240 S.)
ESR-Spektroskopie
(R. Kirmse u. J. Stach, 1985, ca. 160 S.)
Messtechnik ionisierender Strahlung
(W. Stolz, 1985, ca. 160 S.)
Umweltbiophysik
(F. Koessler, 1984, 310 S.)
Pflanzenoekologie Teil I
(G. Lerch, 1985, ca. 210 S., 4. Auflg.)

VEB Deutscher Verlag für Grundstoffindustrie, Leipzig
ABC Umweltschutz
(Autorenkollektiv, 1984, 352 S.)
Geschichte der Elektrochemie
(L. Dunsch, 1985, ca. 176 S.)
Die Geburt des friedlichen Atoms
(D.I. Blochinzew, 1985, ca. 96 S.)
Oberflaechenbehandlung und metallische Beschichtung
(P. Maass u. P. Peissker, 1985, ca. 192 S.)
Werkstoffe der Halbleitertechnik
(Autorenkollektiv, 1985, ca. 344 S.)
Reaktionen an Polymeren
(M. Fedtke, 1985, ca. 264 S.)
Chemische Reaktoren, Ausruestungen und ihre Berechnungen
(Autorenkollektiv, 19850 ca. 504 S.)

IAEA - Publications Catalog
(1985, pp. 285, Division of Publications, IAEA, A-1400 Vienna, Austria, Wagramerstr. 5, P.O.Box 100)

Oesterreichisches Forschungszentrum Seibersdorf GmbH
Abridged Version - G.W. Schwach, Dez. 1984, OEFZS Ber. No. 4302
Auslaugversuche an mit Zement konditionierten radioaktiven Ionenaustauschern - P. Patek, Dez. 1984, OEFZS Ber. No. 4305
ELBEST-Roentgenfluoreszenz-Messanlage fuer die direkte Truebestromanalyse in der Flotation - D. Donhoffer, Jaenner 1985, OEFZS Ber. No. 4306
DNA-Metabolismus und Poly(ADR-Ribose)-Synthese bei beruflich strahlenexponierten Personen - W. Klein u.a., Jaenner 1985, OEFZS Ber. No. 4307
In-Service Supervision of a Prestressed Concrete Pressure Vessel - H. Zemann u.a., Jaenner 1985, OEFZS Ber. No. 4308
Berechnung der Frostsicherheit von wassergefuellten Waermespeichern in Kellerraeumen unter hochalpinen Wetterbedingungen - M. Suda and H. Hick, Jaenner 1985, OEFZS Ber. No. 4309
Online Measurement of the Grain Size Distribution of Sand 0-4 mm Used for the Production of Concrete - Th. Eder u.a., Jaenner 1985, OEFZS Ber. No. 4310

Veranstaltungskalender – Tagungen – Lehrgänge / Calendar – Meetings – Courses

In den Hallen der Schweizer Mustermesse in Basel finden folgende Fachmessen statt:

INELTEC 85

12. Internationale Fachmesse fuer Elektronik und Elektrotechnik vom 10. bis 14. 9. 1985

Einer der wichtigsten Wirtschaftszweige, der vom Maschinenbau ueber die chemische Verfahrenstechnik bis zur Forschung kaum moch einen Bereich kennt, der nicht ohne ein Mindestmass an Elektronik auskommt, hat seinen bewaehrten Treffpunkt: die Ineltec. Nach 2 Jahren besteht nun wiederum die Gelegenheit, sich ueber den aktuellen Stand der Dinge und die Innovationen in saemtlichen Bereichen von Produktion und Dienstleistungen zu informieren - bei ueber 700 Ausstellern aus 25 Laendern, die in sieben uebersichtlich gegliederten Fachgruppen aufgeteilt sind.

SWISSDATA 85

Zum 4. Mal Fachmesse fuer Datenverarbeitung in Industrie, Technik und Forschung vom 10. bis 14. 9. 1985

Dass die Swissdata innerhalb so kurzer Zeit bereits zum vierten Mal veranstaltet wird, beweist zum einen nachdruecklich, dass die Hard- und Softwarenbranche eine typische Wachstumsindustrie ist..
Auskuenfte: Sekretariat Ineltec bzw. Swissdata, Postfach, CH-4021 Basel/Schweiz

T a g u n g:

VIIth Symposium on the Physics and Chemistry of Ice - Grenoble/France, vom 1 - 5 Sept. 1986
Laboratoire de Glaciologie, B.P. 68, F-38402 Saint-Martin-d'Heres Cedex

Ausgewählte Beiträge / Recently published papers

ATOMNAJA ENERGIJA 58(1985)5 (in Russian)

On the 40th Anniversary of the Victory Day

A Comparison of Optimized Models of Nuclear Power Growth - V.P. Brailov et al.

System Principles for Means of Industrial Control of APS Fuel Elements - A.S. Shtan et al.

Formation of Intense Electron Beams of Microsecond Duration - O.A. Gusev et al.

A Pulsed Supply System with Accumulator Storage of Energy - V. Ya. Lomakin et al.

Economic Effectiveness of Radionuclide Neutron Sources - E.V. Kirillov

Variational Methods of Synthesis and Their Application to Neutron-Physical Investigations of Nuclear Reactors (a reviews) - V.V. Khromov et al.

Determination of Some Metal Impurities in Ditolylmethane Coolant - V.A. Ermakov et al.

A Multichannel Scanning Device - A.M. Bobomolov et al.

An Improved Determination of the Reaction Cross-Sections of (n,α_0), (n,α_1), (n,α_2), (n,α_3), (n,α_4), on ^{28}Si at a Neutron Energy of 14.1 MeV - L.I. Klochkova et al.

The Dependence of the Viscosity of Ditolylmethane Organic Coolant on the Products of Its Radiopyrolysis - A.I. Gavrilin et al.

A Position-Sensitive Detector of Thermal Neutrons - E.A. Muminov et al.

A Numerical-Analytic Algorithm for Calculating Radiation Intensity in a Heterogeneous Plane Medium - V.I. Bilenko

ATOMWIRTSCHAFT/ATOMTECHNIK 30(1985)3

ATW-Schnellstatistik:
Kernkraftwerke 1984 - Weltuebersicht

Die zweite nukleare Aere - W. Marth

Schwerpunktthema:
Service im Kernbrennstoffkreislauf

Beschaffung und Vermittlung von Kernbrennstoffen - J. Hoffmann u.a.

Die kundenorientierte Brennelement-Einsatzplanung - R. Schwarzwaelder u.a.

Messung der Oxidschichtdicke an Zry-Huellrohren und -Strukturteilen - W.-R. Kerwath u.a.

Inspektion von LWR-Brennelementen und Kernbauteilen als Wartungsmassnahme - F. Schlemmer u.a.

US-Pruefung von Brennelementen des BE-Wechsels - K. Knecht u.a.

Verdichtung abgebrannter Brennelemente aus Leichtwasserreaktoren - J. Banck u.a.

Planungsarbeiten fuer den Abtransport bestrahlter Brennelemente - P. Schmidt

Sicherstellung unterkritischer Auslegung von Wiederaufarbeitungsanlagen - S. Dagbjartsson u.a.

Tagungsbericht:
Status und Perspektiven der LWR-Sicherheitsforschung (5th International Meeting on Thermal Nuclear Reactor Safety in Karlsruhe)

ATOMWIRTSCHAFT/ATOMTECHNIK 30(1985)2

Die Errichtung des Kernkraftwerks Philippsburg Block 2 - H.-J. Clasen u.a.

Aufgaben und Aufbau des Reaktorschutzsystems - H.-G. Krause u.a.

Leittechnik in der Reaktortechnik - G. Doerrschuck

Abtrennung von Transuranelementen und Spaltprodukten - K. Gompper

Waldschaeden und Kernkraftwerke - M. Hillerbrand u.a.

Umweltaspekte der Energieumsetzung - P. Beck

J. RADIOANAL. NUCL. CHEM. LETT. 94(1985)1

Regularities in gas formation at γ-radiolysis of coals and petroleum residues - I.I. Mustafaev et al.

Substoichiometric radiochemical determination of cadmium with PAN - K. Ashok Rao et al.

Determination of isotopic composition of boron - K.L. Ramakumar et al.

Fast atom bombardment mass spectrometry of pertechnetate - D.D. Fetterolf et al.

J. RADIOANAL. NUCL. CHEM. LETT. 93(1985)6

Monitoring and abatement of environmental mercury pollution using human hair as absorbant - S.S. Krishnan et al.

Production of F-18 with an O-18 enriched water target - I. Huszar et al.

Further attempts to isolate superheavy elements in the meteorite Allende - T. Lung et al.

Determination of Al, As, Co, Cr, Cs, Fe, Mn, Sb, Sc, W and Zn in the workroom air by INAA - E. Haddad et al.

Isotopenpraxis 21 (1985) 6, S. 211 bis 213

Gamma-Radiolyse von 18-Krone-6-ethern

H. Langguth[1]), *E.-J. Langrock*[2])

([1]) Zentralinstitut für Isotopen- und Strahlenforschung der AdW der DDR Leipzig; [2]) Technische Hochschule Leipzig, Sektion Naturwissenschaften)

Der Einsatz von Kronenethern für selektive Trennungen ist in wenigen Fällen gut bekannt. Es wurden Experimente zur Stabilität der Kronenether gegen Gamma-Strahlung durchgeführt. Untersucht wurden 18-Krone-6-ether. Die Bestrahlung erfolgte im Dosisbereich von 5 kGy bis 1 MGy unter Nutzung der ^{60}Co-Gamma-Strahlung. Durch Anwendung der Gaschromatographie und der H-NMR-Spektroskopie konnten die Radiolyseprodukte identifiziert werden. Mit diesen Untersuchungen wird gezeigt, daß die Kronenether für Trennungen in verschiedenen radiochemischen Prozessen nicht nur im Laboratorium sondern auch im technischen Bereich Anwendung finden können.

The ability of crown-ethers for selective separation in a few cases is well known. Initial experiments were made to investigate the stability against gamma-irradiation. In these experiments 18-crown-6-ethers were used. Irradiation was carried out in the dose range from 5 kGy up to 1 MGy using a ^{60}Co-source. Radiolysis products were analyzed by gas chromatography and NMR-spectroscopy. It has been shown that crown-ethers can be used for separation steps in various radiochemical processes not only in a laboratory but also in a technical scale.

Keywords

ethers; gamma radiation; G value; radiation chemistry; radiochemistry; radiolysis

Einleitung

Ziel der vorliegenden Arbeit ist es, an Hand von Strahlenresistenzuntersuchungen die Stabilität von 18-Krone-6-ethern zu bestimmen, um ihre Anwendbarkeit unter den Bedingungen eines radioaktiven Umgebungsniveaus zu beurteilen [1]. Hierzu machte es sich erforderlich, eine Reihe experimenteller Untersuchungen durchzuführen, da in der Literatur keine Vergleichswerte vorzufinden waren. Zu diesem Zweck wurde 18-Krone-6-ether nach erfolgter Strahlenexposition mit verschiedenen Methoden auf radiolytische Zersetzungsprodukte untersucht und das Abbauverhalten der substituierten 18-Krone-6-ether durch Analogieschlüsse [2] charakterisiert.

Experimenteller Teil

Die Bestrahlungsexperimente wurden an der Kobalt-60-Gamma-Bestrahlungsanlage vom Typ „GB 77" bei einer Dosisleistung von 3 kGy/h im Dosisbereich von 5 kGy bis 1 MGy im Temperaturbereich von 20 bis 40 °C durchgeführt.

Zur Bestrahlung wurde der Ether in C_6D_6 gelöst. Das erlaubte nachfolgend die zweifelsfreie Zuordnung aller gasförmigen Protonen enthaltenden Kohlenwasserstoffverbindungen zu den Radiolyseprodukten.

Die Bestrahlung der Proben erfolgte in abgeschlossenen Glasampullen unter Luft. Für die analytische Bestimmung der gasförmigen Radiolyseprodukte sowie der Ausgangssubstanz wurden verschiedene Meßmethoden genutzt.

Gaschromatographie

Die gaschromatographische Bestimmung der Spaltgase erfolgte an gepackten Säulen mit Wärmeleitfähigkeits- und Flammenionisationsdetektoren am Gaschromatographen des Typs 18.3-4 der Firma „Chromatron".

[1]) Anschrift: 7050 Leipzig, Permoserstr. 15

Der Nachweis und die quantitative Erfassung der Radiolysegase H_2, CH_4, C_2H_6 und C_2H_4 wurde unter Anwendung der Trennsäulen Molekularsieb 5A (H_2) und Porapak Q-S (Kohlenwasserstoffe) durch Gasdosierung über eine Gas-Probenschleife durchgeführt. Die Berechnung der Konzentration an radiolytischen Spaltgasen wurde über Absoluteichung mit Wasserstoff und Methan und EDV-Auswertung der Peakflächen vorgenommen.

Protonenkernresonanz

Die kernresonanzspektroskopische Vermessung der protonenhaltigen CH_2-Gruppen im 18-Krone-6-ether erfolgte bei Raumtemperatur am 80-MHz-Kernresonanzspektrometer vom Typ BS 487 C (Tesla) an 0,1 molaren Lösungen von 18-Krone-6-ether in C_6D_6 unter Zusatz von Tetramethylsilan. Für die Berechnung der Protonensignale im bestrahlten 18-Krone-6-ether wurden die integrierten Flächen der Resonanzsignale der CH_2-Gruppen auf die CH-Gruppen von Tetrabromethan als innerem Standard bezogen.

Ergebnisse und Diskussion

Die durchgeführten Untersuchungen zeigen, daß der 18-Krone-6-ether durch den Einfluß von Gammastrahlung radiolytisch abgebaut wird (Tab. 1 u. 2). Im Radiolysegas konnten mittels Gaschromatographie ab Dosen von 10 kGy Wasserstoff und Methan sowie ab Dosen von 100 kGy geringe Mengen an Ethan und Ethen nachgewiesen werden. Die Abspaltung von Wasserstoff nimmt dabei, wie in Abb. 1 gezeigt wird, fast linear mit steigender Dosis zu.

Die NMR-Untersuchungen bestätigen diesen Abbau und zeigen die fast lineare Abnahme der Protonenkonzentration der CH_2-Gruppen. Die Abnahme der Intensität der CH_2-Gruppierung im ppm-Bereich von 3,5 bei einer Bestrahlungsdosis von 1 MGy beträgt 14%.

Tab. 1. Gamma-Bestrahlung von 18-Krone-6-ether in C_6D_6 (gasförmige Radiolyseprodukte)

Probe	Dosis	Dosis-leistung	Ausbeute H_2	*G*-Wert	Ausbeute CH_4	*G*-Wert	Ausbeute C_2H_6/C_2H_4	
	kGy	kGy/h	Mol-% $\times 10^{-2}$		Mol-% $\times 10^{-2}$		Mol-% $\times 10^{-2}$	*G*-Wert
1	10	3	0,085	0,008	0,04	0,004	—	—
2	50	3	0,455	0,009	0,10	0,002	0,039	0,007
3	100	3	1,033	0,012	0,20	0,002	0,069	0,006
4	200	3	1,611	0,009	0,49	0,002	0,143	0,006
5	300	3	2,511	0,009	0,60	0,002	0,180	0,005
6	500	3	3,940	0,008	0,11	0,002	0,213	0,004
7	1000	3	7,920	0,003	2,60	0,010	0,426	0,004

Tab. 2. Gamma-Bestrahlung von 18-Krone-6-ether in C_6D_6 (Abbaurate-Festsubstanz)
Ausgangskonzentration 0,435 Mol/l

Probe	Dosis	Dosis-leistung	Kronen-ether	Differenz	Abbau-rate
	kGy	kGy/h	Mol/l	Mol/l	*G*-Wert
1	10	3	0,4342	0,0012	1,15
2	50	3	0,4271	0,0083	1,60
3	100	3	0,4187	0,0167	1,25
4	200	3	0,4017	0,0337	1,62
5	300	3	0,3912	0,0442	1,42
6	500	3	0,3784	0,0570	1,10
7	1000	3	0,3742	0,0612	0,60

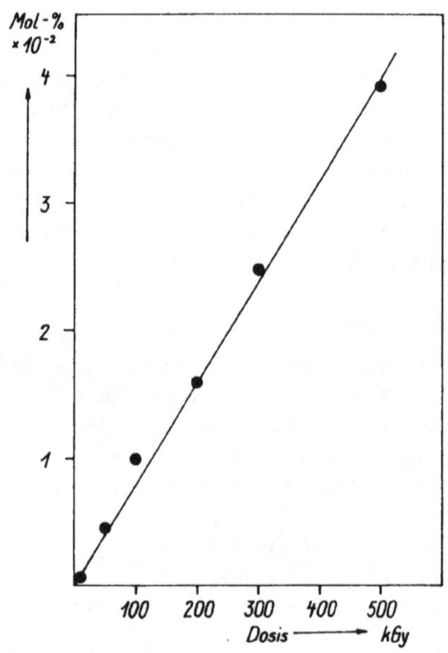

Abb. 1. Ausbeute an Wasserstoff als Funktion der Dosis von 18-Krone-6-ether in C_6D_6

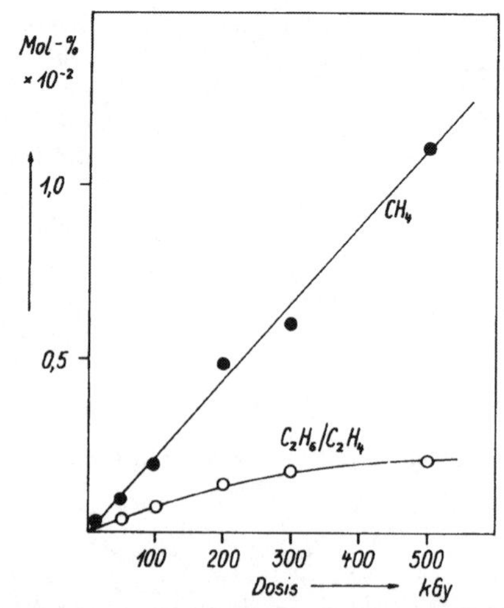

Abb. 2. Ausbeute an Methan und Ethan/Ethen als Funktion der Dosis von 18-Krone-6-ether

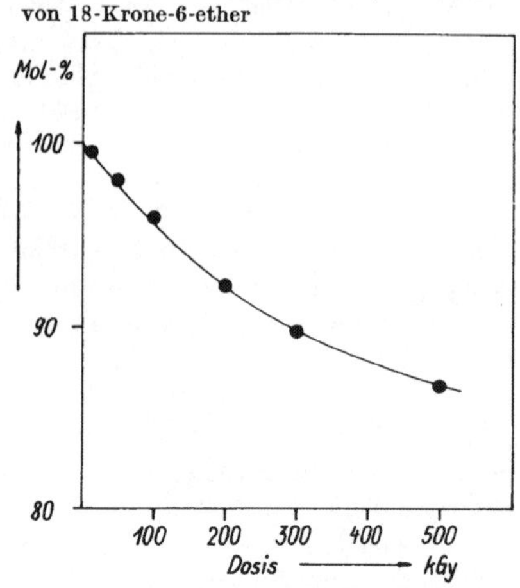

Abb. 3. Abbaurate als Funktion der Dosis von 18-Krone-6-ether in C_6D_6

Wie in Abb. 2 dargestellt, erfolgt die Methanabspaltung linear mit steigender Dosis und nimmt ab 0,5 MGy stark zu. Die Bildung von Ethan und Ethen ist erst ab 0,1 MGy quantitativ nachweisbar und steigt mit Zunahme der Dosis ebenfalls an. Sie ist auf einen C—C-Bindungsbruch im 18-Krone-6-ether sowie auf den Zusammenschluß von CH_3-Radikalen zurückzuführen. Aus der absorbierten Dosis und der Konzentration an Spaltprodukten lassen sich folgende strahlenchemische Ausbeuten für die Spaltgase berechnen:

$G(H_2) = 0{,}009 \pm 0{,}002$
$G(CH_4) = 0{,}002 \pm 0{,}001$
$G(C_2H_6); (C_2H_4) = 0{,}005 \pm 0{,}001$

Für den strahlenchemischen Abbau des reinen 18-Krone-6-ethers konnte unter den gegebenen Bedingungen ein *G*-Wert von $\approx 1{,}25$ ermittelt werden. Das entspricht einer Abbaurate von ca. 15% für eine Dosis von 1 MGy (Abb. 3).

Auf Grund von Analogieschlüssen läßt sich sagen, daß für die vergleichsweise ebenfalls sehr interessanten substituierten 18-Krone-6-ether unterschiedliche Abbauraten zu erwarten sind. So wird die Dicyclohexyl-Verbindung [3, 4] um den Faktor 10 schneller abgebaut, als es bei der Dibenzo-Verbindung (*G*-Wert 0,5) der Fall ist [5, 6]. Damit ergibt sich, daß die Strahlenstabilität in der Reihenfolge

Dibenzo-18-Krone-6 > 18-Krone-6 > Dicyclohexyl-18-Krone-6

abnimmt.

Schlußfolgerungen

Die vorliegenden Ergebnisse zum radiolytischen Abbauverhalten von 18-Krone-6-ethern stellen für die Beurteilung von Anwendungsmöglichkeiten des Komplexbildners zur selektiven Abtrennung von Radionukliden wichtige Aussagen dar. Die durchgeführten strahlenchemischen Untersuchungen ergaben, daß der 18-Krone-6-ether eine relativ strahlenresistente Substanz gegenüber Gammastrahlung im Bereich bis zu Dosen von etwa 200 kGy ist. Eine weitere Steigerung der Bestrahlungsdosis bis 1 MGy führt zum nur teilweisen Abbau des Ethers.

Daraus kann abgeleitet werden, daß für eine Reihe adiochemischer Trennaufgaben Kronenether als selektives Trennungsmittel eingesetzt werden können, ohne daß hinsichtlich ihrer Strahlenresistenz besondere Bedenken bestehen.

Eingegangen am 20. 5. 1984

Literatur

[1] *H. Langguth, J. Schmidt, U. Försterling,* Isotopenpraxis **19** (1983) 000
[2] *G. Földiak,* Radiation Chemistry of Hydrocarbons. Budapest: Akademiai Kiado 1981
[3] *G. R. Freemann, E. D. Stover,* Can. J. Chem. **46** (1968) 3235
[4] *M. G. Robinson, G. R. Freemann,* J. Chem. Phys. **48** (1968) 983
[5] *D. L. Horrocks,* Chem. Phys. **52** (1970) 1566
[6] *V. A. Kronganz,* Int. J. Radiat., Phys. Chem. **1** (1969) 465

Isotopenpraxis **21** (1985) 6, S. 213 bis 216

Radiolysis of Aqueous Solution of Di- and Trivalent Cobalt-1,2-Diaminocyclohexane-N,N,N',N'-Tetraacetic Acid Chelates

M. B. Hafez[1]), *A. G. Abdel-Rehiem*[2])

(Hot Laboratory Center, Atomic Energy Establishment, Cairo, Egypt.)

Gamma radiolysis of aerated and deaerated solutions of di- and trivalent cobalt and 1,2-diaminocyclohexane-N,N,N',N'-tetraacetic acid (DCTA) was investigated. A radiolytic mechanism has been proposed and discussed. It has been shown that the degradation of the chelate is due to the OH radical while the oxidation of the divalent chelate to the trivalent chelate is due to H_2O_2.

Die Gamma-Radiolyse von belüfteten und entlüfteten Lösungen von zwei- und dreiwertigem Kobalt und 1,2-Diaminocyclohexan-N,N,N',N'-tetraessigsäure (DCTA) wurde untersucht. Ein Radiolysemechanismus wird vorgestellt und diskutiert. Es wird gezeigt, daß die Degradation des Chelats auf das OH-Radikal zurückzuführen ist, während die Oxydation des zweiwertigen Chelats zum dreiwertigen Chelat durch das H_2O_2 bewirkt wird.

Keywords

chelates; cobalt compounds; gamma radiation; G value; irradiation; radiation chemistry; radiolysis

Introduction

Radiolysis of metal chelates in aqueous solutions has been the interest of many authors [1, 2]. It has been reported that OH, the oxidising radical which is formed during radiolysis, attacks the ligand and the hydrated electron, the principal reducing spieces, attacks the central metal ion.

1,2-Diaminocyclohexane-N,N,N'N'-tetraacetic acid (DCTA) is known as a strong chelating agent. The formation of the soluble 1:1 (Co(II)-DCTA) system in *p*H range 5·5 to 8·5 was determined spectrophotometrically [3]. The 1:1 chelate between Co(III) and (DCTA) was detected in the *p*H range 2·5 to 8·5.

The present work illustrates the radiation characteristics of the (Co(II)-DCTA) and (Co(III)-DCTA) complexes in the presence, and absence, of oxygen.

Experimental

All chemicals used were of analytical grade. DCTA was purified several times from ether. Tridistilled water was used for preparing all solutions. The (Co(II)-DCTA) complex was prepared by mixing a known volume of 8×10^{-3} M Co(III) with the same volume of 8×10^{-3} M, $1{\cdot}6 \times 10^{-2}$ M or 4×10^{-2} M (DCTA). The *p*H of the solutions was then adjusted to a known value using NaOH-free carbonate. The (Co(III) + DCTA) complex was prepared by oxidising the (Co(II) + DCTA) complex at *p*H = 8·5 with few drops of H_2O_2 [4]. Decomposition of (Co(II) + DCTA) was done using 8×10^{-2} M Co(II).

Irradiations were carried out using γ rays from ^{60}Co at a dose rate $1{\cdot}666 \times 10^3$ rd/min. The dose rate was determined by Frick ferrous sulfate dosimeter.

All optical measurements were carried out using Prolabo spectrophotometer. The *p*H of solutions were done using Pye Unicam *p*H meter type 292. The hydrogen peroxide was determined with titanium sulfate reagent [5]. Formaldehyde was determined with chromotropic acid [6].

Results and Discussion

1. *Aerated Solutions*

Fig. 1. shows the absorption spectra of; 4×10^{-2} M Co(II), a mixture of 8×10^{-2} M Co(II) and 8×10^{-2} M (DCTA) at *p*H 5·5, and a mixture of 8×10^{-3} M Co(III) and 8×10^{-3} M (DCTA) at *p*H 3·5. The characteristic wavelengthes

[1]) All correspondences should be made to: Prof. Dr. *M. B. Hafez,* Chemistry Department, UAE University, AL-AIN, P. O. Box 15551, United Arab Emirates
[2]) Chemistry Department, Faculty of Science, UAE University

Tab. 1. Characteristics of Co(II); Co(III) and their complexes

Ion	max^nm	M^{-1} cm^{-1} l	Compound	Mole ratio		
Co^{2+}	500	6—	(Co^{+2} + DCTA)	1:1	550	10
Co^{+3}	—	—	(Co^{+3} + DCTA)	1:1	515	100

Tab. 2. Experimental yield *G* values obtained during radiolysis of (Co(II) + DCTA) and (Co(III) + DCTA) chelates

Condition	Mole ratio Co:DCTA	Oxidation yield Co(III) = 8×10^{-3} M *p*H 2·5	Decomposition yield *p*H 4·5	
			Co(II) + DCTA Co(II) 4×10^{-2} M	Co(III) + DCTA Co(III) = 4×10^{-3} M
Aerated	1:1	1·51	2·71	5·64
Solution	1:2	3·12	2·63	5·51
Deaerated	1:1	—	1·41	4·42
Solution	1:2	—	1·42	4·38

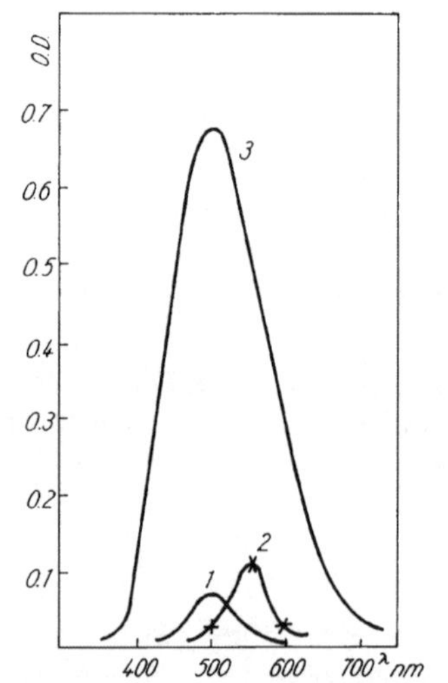

Fig. 1. Absorptions spectra of: (1) 4×10^{-2} M Co(II), (2) 4×10^{-2} M Co(II) + 4×10^{-2} M (DCTA) at *p*H 3·5 and (3) 4×10^{-3} M Co(III) + 4×10^{-3} M DCTA at *p*H 2·5

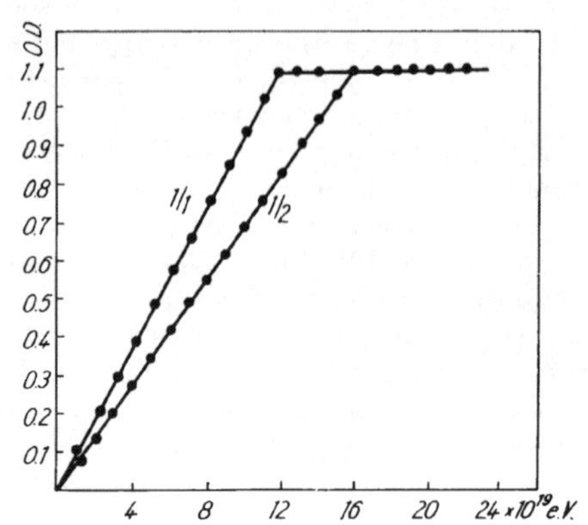

Fig. 2. Gamma-radiation-induced oxidation of (Co(II) + DCTA) at (Co(II))/(DCTA) = 1/1 and 1/2 at different *p*H values

1 — *p*H = 1·5; 2 — *p*H = 2·5; 3 — *p*H = 3·5; 4 — *p*H = 5·5; 5 — *p*H = 7·5; Optical measurement at 515 nm

and the extinction co-efficient of each compound are given in Tab. 1.

Experimental results showed that, at *p*H intervals from 2·5 to 5·5, divalent cobalt in presence of DCTA was oxidised during radiolysis to (Co(III) + DCTA). Fig. 2 shows the radiolytic oxidation of (Co(II) + DCTA) in presence of DCTA to (Co(III) + DCTA) at different *p*H values as a function of absorbed dose. The radiolytic oxidation yield was calculated (Tab. 2) and was found decrease with the *p*H and with the DCTA concentration.

Obtained results showed also that the (Co(II) + DCTA) and (Co(III) + DCTA) complexes were decomposed, during radiolysis. As the DCTA concentration increased, decomposition of (Co(II) + DCTA) and (Co(III) + DCTA) complexes decreased. Fig. 3 and 4 show that the radiolytic decomposition of (Co(II) + DCTA) and (Co(III) + DCTA) complexes at *p*H 5·5 respectively, are all linear with the absorbed dose. The *G* value for each case was calculated and the obtained values are given in Tab. 2.

a) *Mechanism of oxidation Co(II) in presence of DCTA by gamma radiation.* The radiation decomposition of water

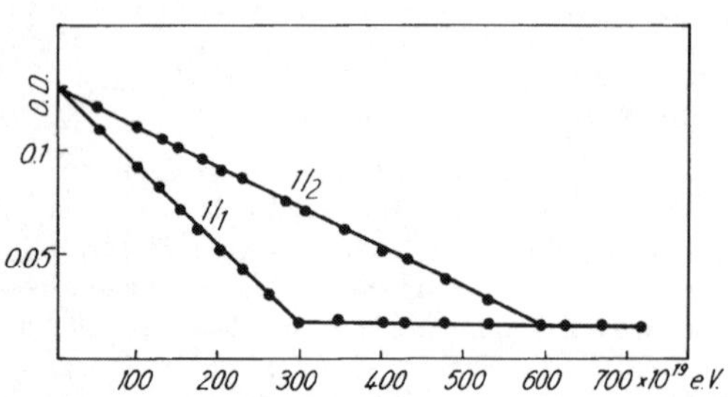

Fig. 3. Gamma-radiation-induced degradation of (Co(II))/(DCTA) = 1/1 and 1/2 chelate at *p*H = 4·5

(Co(II)) = 4×10^{-2} M; Optical measurements at 515 nm

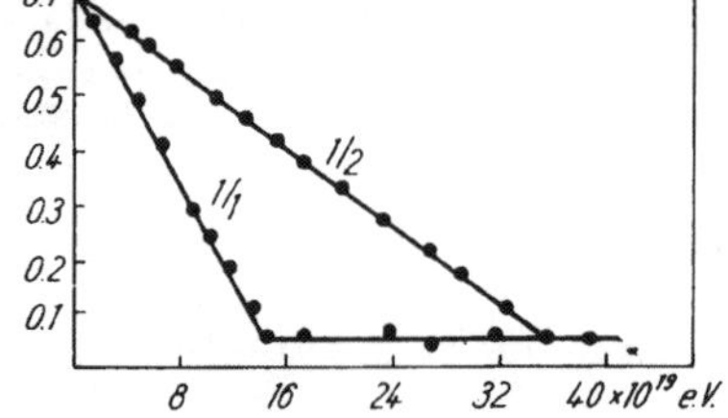

Fig. 4. Gamma-radiation-induced degradation of (Co(III))/(DCTA) = 1/1 and 1/2 chelate at *p*H 2·5

(Co(III)) = 4×10^{-3} M; Optical measurement at 525 nm is presented by the following radicals and molecules; OH, e^-_{aq}, H, H_2, H_2O_2 the G value of each is now well established [7].

As the radiolysis is carried out in presence of oxygen air, the species H and e^-_{aq} react with oxygen. Their reaction with oxygen are more faster than with metal chelate [2]

$$H + O_2 = HO_2 \quad (1)$$

$$e^-_{aq} + O_2 = O_2^- \quad (2)$$

The radicals HO_2 und O_2^- may either react with each other or react with the metal ion. The former possibility seems to occur.

$$O_2^- + O_2^- \xrightarrow{2H_2O} H_2O_2 + O_2 + OH^- \quad (3)$$

On the other hand, divalent cobalt, in presence of DCTA, at pH 2·5 exists as cation state i.e. not in complex form, while trivalent cobalt forms a complex with DCTA at $p\text{H} \geqq 2{\cdot}0$ [3]. It is possible to assume that the radiolytic species OH attacks the cobaltus ion itself rather than the ligand to produce cobaltic ion which can form (Co(III) + DCTA) complex with the non degradated DCTA.

$$Co^{2+} + OH = Co^{+3} + OH^- \quad (4)$$

Since the oxidation rate of cobaltus ion by H_2O_2 is effective [7], the oxidation yield may be expressed by:

$$\therefore G(Co^{+3}) = G(OH) + G(H_2O_2) \quad (5)$$

The obtained value (Tab. 2) is in good agreement with the above calculated value.

b) *Mechanism of degradation of (Co(II) + DCTA) and (Co(III) + DCTA) complexes by gamma radiation.* It has earlier been stated, in case of radiolysis of metal chelate compounds [8], that the OH radical attacks the chelating agent of the metal chelate, leading to the abstraction of hydrogen atom from the carbon of the COOH group. The following reaction may then be considered

$$Co^{+2}RNCH_2COO^- + OH = Co^{+2}RNCHCOO^- + H_2O \quad (6)$$

$$Co^{+3}RNCH_2COO^- + OH = Co^{+3}RNCHCOO^- + H_2O \quad (7)$$

where $Co^{+2}RNCH_2COO^-$ and $Co^{+3}RNCH_2COO^-$ represent the (Co(II) + DCTA) and (Co(III) + DCTA) respectively.

In presence of oxygen, the produced radicals react with oxygen

$$Co^{+2}RNCHCOO^- + O_2 = Co^{+2}RNCH(O_2)\,COO^- \quad (8)$$

$$Co^{+3}RNCHCOO^- + O_2 = Co^{+3}RNCH(O_2)\,COO^- \quad (9)$$

The oxygenated radicals are unstable they may decompose as

$$Co^{+2}RNCH(O_2)COO^- \xrightarrow{+H_2O} Co^{+2}RNH + CH_2O + O_2^- + CO_2 \quad (10)$$

$$Co^{+3}RNCH(O_2)COO^- \xrightarrow{+H_2O} Co^{-3}RNH + CH_2O + O_2^- + CO_2 \quad (11)$$

Formaldehyde was detected and its $G(CH_2O)$ was found to be equal to $0{\cdot}5 \pm 0{\cdot}1$.

It is apparent from above equations that the decomposition of (Co(II) + DCTA) and (Co(III) + DCTA) complexes must equivalent to G(OH). Actually the observed decomposition value for (Co(II) + DCTA) complex is not far from that expected, while the obtained value for (Co(III) + DCTA) is much higher than that expected. Therefore an additional reaction may occur in case of decomposition of (Co(III) + DCTA) complex.

The hydrogen peroxide yield may react with (Co(II) + DCTA) chelate. Its action can be represented by:

$$Co^{+2}RNCH_2COO^- + H_2O_2 = Co^{+3}NCH_2COO^- + OH + OH^- \quad (12)$$

On the other hand, the H and e^-_{aq} react with the central atom and reduce (Co(III) + DCTA) complex according to:

$$Co^{+3}RNCH_2COO^- + HO_2 = Co^{+2}RNCH_2COO + O_2 + H^+ \quad (13)$$

The reduction of (Co(III) + DCTA) complex should be equivalent to ($G\,e^-_{aq} + G_H$), while the oxidation of (Co(II) + DCTA) should be equivalent to $G_{H_2O_2}$. By comparing these values, and since a decrease in the optical density at 515 nm means a reduction of (Co(III) + DCTA) complex also its degradation, it can be postulated that reduction of (Co(III) + DCTA) complex by γ rays, should occur simultaneously with its degradation. The G value of the produced Co(II) after degradation and reduction of (Co(III) + DCTA) complex should be equal to ($G_{OH} + G_{HO_2}$). The obtained value (Tab. 2) is not far from that expected.

2. *Deaerated Solutions*

Obtained results showed that Co(II), in presence of DCTA, at 3·5 pH 5·5 were slightly oxidised during radiolysis. In deaerated solution, the OH and H_2O_2 have the same action as in aerated solution, while H and e^-_{aq} reduce an equivalent of Co(III). Therefore the trivalent cobalt expected yield at pH 3·5 should be equal to:

$$G(Co(III)) = G(OH) - G(H + e^-_{aq}) \quad (14)$$

It is clear that Co(III) does not produce by radiolysis, which is confirmed experimentally.

Results showed also that degradation of (Co(II) + DCTA) and (Co(III) + DCTA) complexes, in deaerated solution, was occurred. As in case of radiolysis of (Co(II) or Co(III) + DCTA) complexes in a aerated solutions, the OH radical attacks the (Co(II) + DCTA) and (Co(III) + DCTA) chelates to form the intermediate radicals ($Co^{+2}RNCH\,COO^-$) and ($Co^{3+}RNCH\,COO^-$) respectively. The obtained decomposition yield values (Tab. 2) are much lower than G(OH). Therefore it can be proposed that these radicals, in the absence of oxygen, may react with each other

$$2\,Co^{+2}RNCH\,COO^- + H_2O = Co^{+2}RNCH_2COO^- + Co^{+3}RNCH(OH)\,COO^- \quad (15)$$

$$2\,Co^{+3}RNCH\,COO^- + H_2O = Co^{+3}RNCH_2COO^- + Co^{+3}RNCH(OH)\,COO^- \quad (16)$$

It is clear that (Co(II) + DCTA) or (Co(III) + DCTA) are regenerated again. Moreover hydrogen atom may reacts also with (Co(II) + DCTA) or (Co(III) + DCTA) complexes

$$Co^{+2}RNCH_2COO^- + H = Co^{+2}RNCHCOO^- + H_2 \quad (17)$$

$$Co^{+3}RNCH_2COO^- + H = Co^{+3}RNCHCOO^- + H_2 \quad (18)$$

$$Co^{+3}RNCH_2COO^- + H = Co^{+2}RNCH_2COO^- + H^+ \quad (19)$$

Therefore the decomposition yield of (Co(II) + DCTA) complex should correspond to a value equivalent to 1/2 $(G_H + G_{OH})$. The observed value of (Co(II) + DCTA) complex is equivalent to that expected (Tab. 2), while the observed value in case of (Co(III) + DCTA) is much higher than that expected, the hydrated electrons in absence of oxygen may reduce the (Co(III) + DCTA).

$$Co^{+3}RNCH_2COO^- + e^-_{aq} = Co^{+2}RNCH_2COO^- \quad (29)$$

while the H_2O_2 may oxidise the (Co(II) + DCTA) complex

$$Co^{+2}RNCH_2COO^- + H_2O_2 = Co^{+3}RNCH_2COO^- + OH^- + OH \quad (21)$$

The hydrogen peroxide again may disapper by the reaction

$$H_2O_2 + H = OH + H_2O \quad (22)$$

Comparing the yield values of H_2O_2 and HO_2 in the above reactions. It can be concluded that hydrated electrons reduce (Co(III) + DCTA) complex simultaneously when OH attacks the chelate and the produced divalent cobalt should be equivalent to

$$G(Co(II)) = 1/2\,(G_{OH} + G_H) + G^-_{eq} \quad (23)$$

the obtained value is equivalent to that expected.

Received March 22, 1984

References

[1] *B. Sramkov, J. Zyka, J. Dolezal*, J. electroanal. Chem. **30** (1977) 1977
[2] *M. B. Hafez*, CEA-R-3521, France 1968
[3] *E. Masri, M. Haissinsky*, Chem. Phys. **60** (1963) 397
[4] *S. N. Bhattacharyya, K. P. Kundu*, Radiation Research **51** (1972) 45
[5] *T. R. Bhat, J. Krishnamurthy*, J. Inorg. Nucl. Chem. **25** (1963) 1147
[6] *B. H. J. Bielski, A. O. Allen*, Intern. H. Radiat. Phys. Chem. **1** (1963) 153
[7] *S. N. Bhattacharyya, K. P. Kundu*, Intern. J. Radiat. Phys. Chem. **3** (1971) 1
[8] *M. Haissinsky, P. Patigny*, J. Chem. Phys. **59** (1962) 675

Isotopenpraxis **21** (1985) 6, S. 216 bis 218

^{99m}Tc Generator — Prevention of Radiolytically Induced Reduction of Elution Yield

J. L. Vučina, S. M. Milenkovič

(Boris Kidrič Institute of Nuclear Sciences, Institute for Radioisotopes, Beograd)[1])

For ^{99m}Tc generator activities ranging from 74—92·5 GBq, an adsorbent consisting of a mixture of alumina and silica gel was modified by copper addition. The quantity of copper adsorbed on alumina was in the range of 10—15 mg Cu/g alumina. It was found that saline solution does not influence the performance of the columns. A high elution yield of ^{99m}Tc and good quality of the eluate was obtained both for the "wet" and "dry" column.

Für ^{99m}Tc-Generator-Aktivitäten von 74 ... 92,5 GBq wurde ein Adsorptionsmittel, bestehend aus einem Gemisch von Tonerde und Silikagel, durch Kupferzusatz modifiziert. Die Menge des von der Tonerde adsorbierten Kupfers lag bei 10 ... 15 mg Cu/g Tonerde. Es wurde festgestellt, daß Salzlösungen den Wirkungsgrad der Säulen nicht beeinflussen. Eine hohe Eluierungsausbeute für ^{99m}Tc und eine gute Qualität des Eluats wurden sowohl bei der „nassen" als auch bei der „trockenen" Säule erzielt.

Keywords

molybdenum 99; radioisotope generators; technetium 99m

Introduction

The elution yield of ^{99m}Tc is a function of the ^{99}Mo activity and self-irradiation dose (caused by the decay of ^{99}Mo) absorbed in the generator. These doses generate highly reactive species. Among them are hydrated electrons which seem to be responsible for the reduction of pertechnetate. It is known that TcO_4^- has high affinity toward electrons [1]. The decrease of the elution yield of ^{99m}Tc caused by the radiationinduced reduction of ^{99m}Tc-pertechnetate could be supressed by the addition of an oxidizing agent which should re-oxidize the reduced ^{99m}Tc. This procedure, however, has drawbacks regarding possible interferences in the subsequent use of eluates.

A more favourable method could be the use of so-called scavengers which, by reaction with hydrated electrons, prevent the reduction of ^{99m}Tc-pertechnetate.

In our previous papers [2, 3] we have discussed the possible use of a modified sorbent for generators of higher activities. These generators must be dried after each elution.

In this paper we report of experiments with copper adsorbed in the generator column. According to data published in the literature [4, 5], copper, as well as some other elements [6], has favourable rate constants for the reaction with hydrated electrons. We have obtained good results for generators of activities up to 74—92·5 GBq for both "dry" and "wet" columns.

Experimental

The detailed generator construction was reported elsewere [7]. Copper was adsorbed on alumina (Woelm N) by static adsorption from a diluted water solution of $CuCl_2 \cdot 2\,H_2O$ [8]. After decanting and washing, the treated alumina was dried at 200 °C for 6 hours.

The upper part of the generator column in which ^{99}Mo will be adsorbed contained 1 g of the mixture of treated alumina and silica gel, so that the ratio of $Al_2O_3 \cdot Cu$ was about 10—15%(wt). Under this layer the column contained 1 g of untreated alumina. Generators containing the same

[1]) Address: 11001 Beograd, P.O. Box 522, Yugoslavia

column composition but without copper were used for comparison.

A known activity of ^{99}Mo was adsorbed and the generators eluted with saline solution every day in the course of ten days. The elution yield of ^{99m}Tc was determined starting from the calculated activity of ^{99m}Tc present in the column at the time of elution. The radionuclidic purity of eluates was checked radiometrically with respect to the content of gamma emitting radionuclides. A multichannel analyser with a Na(I) detector (Typ Zoomax, Sein) and 4096 channel (Sein) analyser with a Ge(Li) detector were used. The radiochemical purity was determined by ascending chromatography on Whatman No. 1 paper in different solvent mixtures. Paper impregnated with quinalizarine was used for the spot test for eventually present aluminium ions. The content of copper was determined by atomic absorption spectrophotometry. The content of H_2O_2 in the eluates was determined by the colorimetric method [9].

Results and Discussion

The results of our experiments with generator columns doped with cupric ion are shown in Tab. 1. It can be seen that in the region of adsorbed ^{99}Mo of about 74 GBq the columns still give high elution yields of about 80%. It is necessary to point out that these data refer to "wet" columns. It means that although saline solution is present in the columns during the growth of ^{99m}Tc, elution yields are not affected. The content of adsorbed copper was found to be in the region of 10—15 mg Cu/g alumina.

Tab. 1. Characteristics of the ^{99m}Tc generators with columns doped with copper
Column: Upper layer 0·5 g of $SiO_2 + Al_2O_3 \times Cu$
Under layer 1 g Al_2O_3 (acid, Woelm R)
Eluence: 0·9% NaCl, 10 ml
^{99}Mo activity adsorbed at the time of first elution: 74—92·5 GBq

Elution[a]) days	^{99m}Tc elution efficiency "wet" columns[b]), %
1	81·5
2	79·3
3	80·7
4	76·3
5	78·4
6	79·3
7	78·2
8	79·0
9	83·0
10	77·7

a) Elutions were done in the intervals of 22—24 hours.
b) After each elution the columns were kept wet (0·9% NaCl).

In Table 2 we have shown the data obtained with columns used for comparison. The adsorbent was the same (mixture of alumina and silica gel) but did not contain copper. In our previous papers we have shown that this adsorbent used in generator columns enables the production of ^{99m}Tc generators with good performance in the broad region of adsorbed ^{99}Mo activities. However, when this activity rises over 55—74 GBq, the elution yields decrease to about 65—70%. This is true only for the "dry" columns. As can be seen in Tab. 2 the presence of saline solution in the columns has a considerable effect causing the decrease of elution yields. This means that in the case of improper elution, the saline solution which remained in the column could have serious consequences on elution next day.

Tab. 2. Elution yields of ^{99m}Tc in ^{99m}Tc generators without copper addition
Column: Upper layer 0·5 g of $SiO_2 + Al_2O_3$
Under layer 1 g of Al_2O_3 (acid, Woelm R)
Eluence: 0·9% NaCl, 10 ml
^{99}Mo adsorbed activity at the time of first elution: 67—81 GBq

Elution[a]) days	Elution yield of ^{99m}Tc	
	"wet" column[b]), %	"dry" column[c]), %
1	20·3	67·5
2	22·2	66·7
3	5·7	68·7
4	12·8	66·8
5	9·9	66·4
6	8·4	66·6
7	25·0	67·2
8	23·6	66·8

a) Elution were done in the intervals of 22—24 hours
b) Between elutions the columns were kept wet (0·9% NaCl)
c) After each elution the columns were dried by passing 40—50 ccm of sterilized air

As can be seen, the use of copper as the scavenger for electrons incorporated in the generator system has a positive effect: the elution yields remain high (about 80%) and such columns are not sensitive to the presence of saline solution. The explanation of the action of copper, which is present in the form of Cu^{2+}, lies in the rapid reduction of the cupric ion by e^-_{aq}.

The characteristics of the ^{99m}Tc eluates were also determined and compared. The obtained data were similar to those obtained and published earlier [10, 11]. All parameters concerning the radionuclidic, radiochemical and chemical purity were within the permitted limits. The content of copper was also determined. It was found that the content of copper which could be found in the eluates did not exceed $1-2 \cdot 10^{-7}$ g Cu/ml of eluate. The content of hydrogen peroxide in the ^{99m}Tc eluate immediately after generator elution was found to be 2—3 µg/ml of eluate.

Conclusion

The proposed adsorbent for ^{99m}Tc generators consists of a mixture of alumina and silica gel. We obtained good results but it was necessary for the columns to be dried after elution. The presence of saline solution caused a decrease of elution yields of ^{99m}Tc. By introducing copper into the mixture, by adsorbing it on alumina, we have improved the performance of the columns. A copper content in the range of 10—15 mg Cu/g alumina enables the maintenance of high elution yields even in the "wet" columns in the region of adsorbed ^{99}Mo activity up to 74—92·5 GBq. The quality of the ^{99m}Tc eluates meets all requirements regarding radionuclidic, radiochemical and chemical purity.

Received May 16, 1984

References

[1] *P. K. Mathur, K. S. Venkatesnarly*, B.A.R.C.-489 (1970)

[2] *J. L. Vučina, Lj. M. Jaćimović, S. M. Milenković*, IV Int. Symp. on Radiopharmaceutical Chemistry, Jülich (FRG) 1982, p 298

[3] *J. L. Vučina, Lj. M. Jaćimović, S. M. Milenković*, J. Radioanal. Chem. **81** (1984) 171

[4] *G. V. Buxton, P. M. Selers*, Coorination Chem. Reviews **22** (1977) 195

[5] *M. T. Nenadović, O. I. Mićić, N. M. Dimitrijević*, Bull. Soc. chim. Beograd (in press)

[6] *R. E. Boyd*, Radiochim. Acta **30** (1982) 123

[7] *S. M. Milenković, J. L. Vučina, Lj. M. Jaćimović, E. S. Karanfilov, T. V. Memefović*, Isotopenpraxis **19** (1982) 85

[8] *O. C. Smith*, Inorganic Chromatography. D. Van Nostrand Comp., Inc. New York (1953)

[9] *D. J. Savage*, The Analyst **76** (1951) 224

[10] *S. M. Milenković, J. L. Vučina, Lj. M. Jaćimović, T. V. Memedović, E. S. Karanfilov*, Radiol. Iugosl. **17** (1983) 181

[11] *V. Jovanović, T. Memedović, T. Maksin, J. Bzenić, Z. Solareić,v N. Terzić*, Radiol. Iugosl. **17** (1983) 171

Isotopenpraxis **21** (1985) 6, S. 218 bis 220

Sterilisation of Al_2O_3 for Production of ^{99m}Tc Generators

Lj. M. Jaćimovic, S. M. Milenkovic

(The Boris Kidrič Institute of Nuclear Sciences, Vinča, Institute for Radioisotopes)[1])

The effect of sterilisation of Al_2O_3 (by radiation, thermal treatment and in the autoclave) on its adsorption capacity and efficiency of elution of ^{99m}Tc has been investigated. No difference has been observed in the adsorption capacity of sterilised and unsterilised Al_2O_3 samples. However, efficiency of elution of ^{99m}Tc differed in sterilised and unsterilised samples, and even depended on the sterilisation procedure used. The highest elution efficiency was found with unsterilised Al_2O_3 (94%), somewhat lower with thermally sterilised (90%) and the lowest with ^{60}Co gamma radiation sterilised Al_2O_3 (85%). Al_2O_3 sterilised at 300 °C has slightly higher adsorption capacity with respect to molybdenum.

Die Auswirkungen der Sterilisierung von Al_2O_3 (durch Bestrahlung, Wärmebehandlung und im Autoklaven) auf dessen Adsorptionsfähigkeit und Wirksamkeit hinsichtlich der Elution von ^{99m}Tc wurden untersucht. Es wurde kein Unterschied in der Adsorptionsfähigkeit von sterilisierten und nichtsterilisierten Al_2O_3-Proben festgestellt. Allerdings differierte die Wirksamkeit hinsichtlich der Elution von ^{99m}Tc bei sterilisierten und nichtsterilisierten Proben; sie hing sogar von dem angewandten Sterilisierungsverfahren ab. Die höchste Elutionswirkung wurde bei nichtsterilisiertem Al_2O_3 (94%) festgestellt, eine etwas niedrigere bei thermisch sterilisiertem Al_2O_3 (90%) und die niedrigste bei durch ^{60}Co-Gammastrahlung sterilisiertem Al_2O_3 (85%). Bei einer Temperatur von 300 °C sterilisiertes Al_2O_3 hat in bezug auf Molybdän eine etwas höhere Adsorptionsfähigkeit.

Keywords

adsorption; aluminium oxides; molybdenum; radioisotope generators; sterilization; technetium 99m

Introduction

It is well known that Al_2O_3 is used as the adsorber in chromatographic column for production of ^{99m}Tc generators. Since capacity for adsorption is one of the most important properties of Al_2O_3, great attention is paid to determination of its adsorption characteristics under conditions existing in the system $^{99}Mo/^{99m}Tc$ generator [1–3].

As the eluate from the ^{99m}Tc generator, used in nuclear medicine as ^{99m}Tc-pertechnetate, must be sterile, sterilisation is necessarily performed in production of these generators. Chromatographic alteration of Al_2O_3 which occurs during sterilisation in some cases resulted in the necessity for investigation of the effect of applied sterilisation procedure.

Al_2O_3 used in regular production of ^{99m}Tc generators was irradiated with ^{60}Co, treated thermally with hot dry air stream or autoclaved. The effect of these procedures on adsorption capacity of Al_2O_3 was investigated with respect to molybdenum.

Experimental

Materials: Used Al_2O_3-R was Woelm, in acid form, grain size 63–200 μm. A glass column with sinter (8 mm in dia.) with lg Al_2O_3 was used. Molybdate solution for adsorption (0·01 mol/dm³ MoO_3 or 0·959 mg Mo/ml) was prepared by dissolving MoO_3 (Merck) in 4 mol/dm³/NaOH; pH was adjusted with diluted HCl. Molybdate was labelled with the fission ^{99}Mo (IRE-Belgium). Radioactivity of the solution was 55·5 kBq/ml. The effect of elution of ^{99m}Tc from the column of Al_2O_3 was investigated using ^{99}Mo with activity of 37 MBq/ml, without carrier. The column was eluted with physiological solution (0·9% NaCl).

Adsorption of ^{99}Mo and elution of ^{99m}Tc

Adsorption was carried at room temperature, with molybdate solution at 3 different pH values: 1·50; 4·00 and 5·50. First 100 ml of bidistilled water were passed through the column containing lg Al_2O_3 acidified to a determined pH. Elution rate was 20 ml/cm² s. Radioactive molybdate solution was passed through the column in 2 ml portions. The eluate was first collected in penicillin bottles and activity of ^{99}Mo was then measured.

Efficiency of elution of ^{99m}Tc was investigated using a glass column containing lg Al_2O_3 into which was introduced 1 ml of ^{99}Mo solution. The column was rinsed with 5 ml of physiological solution. Technetium was eluted with the same solution first after 24 h and then on each following day. Activity measurements of ^{99}Mo and ^{99m}Tc were performed immediately after elution.

[1]) Address: 11001 Beograd, P.O. Box 522, Yugoslavia

Measurement of ^{99}Mo and ^{99m}Tc radioactivities

Activity of ^{99}Mo was measured on an automatic ICN gamma counter (Belgium), previously adjusted to measure energies between 700–820 keV, in order to eliminate radiation effect of the always present ^{99m}Tc. The values, expressed in imp/min, mostly between 10·000 and 40·000, were compared with those of the stock solution so that the amount of adsorbed molybdenum was calculated from the difference between these activities.

Eluted ^{99m}Tc was measured in an ionisation chamber (Mediac) while efficiency of elution was calculated by comparing measured activities with the theoretical ^{99m}Tc activity. The content of ^{99}Mo was controlled in each eluate. Breakthrough of molybdenum has not been observed in any case and for this reason these results are not presented.

Sterilisation of Al_2O_3

Al_2O_3 was sterilised in three way: by irradiation with ^{60}Co, thermally, in hot dry air stream, and in an autoclave [4]. Sterilisation with gamma radiation was performed using a radiation unit of 37×10^5 GBq from the Laboratory of Solid State Physics and Radiation Chemistry at Vinča. Irradiation dose was $2{\cdot}5 \times 10^4$ Gy. Thermal sterilisation was carried out in hot dry air stream, for 2 and 22 h at 150 °C and for 2 h at 300 °C. Sterilisation in the autoclave was performed for 30 min at 120 °C.

Sterility of samples was tested by direct inoculation of corresponding nutrient substrates, for 7 days [5]. All Al_2O_3 samples were sterile.

Results and Discussion

The results on investigation of adsorption characteristics of Al_2O_3 with respect to molybdenum are presented in Tab. 1. Breakthrough and saturation capacity of 5 types of Al_2O_3 samples, at 3 different pH values, have been determined. Four samples were sterilised: the first, previously irradiated with ^{60}Co was heated at 150 °C for 2 h, the second was only irradiated with ^{60}Co, the third – only heated at 150 °C for 2 h and the fourth – heated at 300 °C for 2h.

By comparing the obtained breakthrough and saturation capacities of sterilised samples for the same pH values with those of unsterilised Al_2O_3, it is evident that these capacities of Al_2O_3 for molybdenum remain unchanged after sterilisation, independent of used sterilisation procedure (radiation or thermal). Al_2O_3 sterilised at 300 °C has slightly increased capacities for molybdenum.

Dependence of Al_2O_3 adsorption capacity on pH has been discussed in our earlier paper [2].

Efficiency of elution of ^{99m}Tc from the columns with unsterilised Al_2O_3 and six Al_2O_3 samples sterilised in different ways is shown in Tab. 2. As it can be seen from the Table, elution efficiency of ^{99m}Tc was the highest with unsterilised Al_2O_3 (94%), somewhat lower with thermally sterilised Al_2O_3 (90%) and the lowest with gamma radiation sterilised Al_2O_3 (85%). It can also be seen in the Tab. 2 that elution efficiency of ^{99m}Tc remained unchanged independent on whether Al_2O_3 had been heated, or not, after irradiation.

By measuring pH of the suspension (lg Al_2O_3 + 10 ml H_2O) no change of pH values of unsterilised and sterilised Al_2O_3 samples has been observed (Tab. 3).

Tab. 3. Dependence of pH of Al_2O_3 water suspension on the sterilisation procedure used

Sample	pH (lg Al_2O_3 + 10 ml H_2O)
Al_2O_3 unsterilised	4·00
Al_2O_3 irradiated ($2{\cdot}5 \cdot 10^4$ Gy) heated 2 h at 150 °C	4·00
Al_2O_3 irradiated ($2{\cdot}5 \cdot 10^4$ Gy)	4·00
Al_2O_3 heated 2 h at 150 °C	4·00
Al_2O_3 heated 22 h at 150 °C	4·00
Al_2O_3 autoclaved 0·5 h at 120 °C	4·00
Al_2O_3 heated 2 h at 300 °C	4·00

Conclusion

Al_2O_3 (Woelm-R) used in regular production of ^{99m}Tc generators was sterilised in 3 ways: by irradiation with ^{60}Co, by thermal heating at 150 °C and 300 °C and by

Tab. 1. Breakthrough (B) and saturation (S) capacities of Al_2O_3 (Woelm-R, acid) for molybdenum in dependence of pH and sterilisation procedure
Column: lg Al_2O_3 (Woelm-R, acid); Flow rate: 20 ml min^{-1} cm^{-2}; Concentration: 0·01 mol Mo/dm^3; Capacities expressed in mg Mo/lg Al_2O_3

pH	Al_2O_3 unsterilised		Al_2O_3 irradiated $2{\cdot}5 \cdot 10^4$ Gy heated 2 h at 150 °C		Al_2O_3 irradiated $2{\cdot}5 \cdot 10^4$ Gy		Al_2O_3 heated 2 h at 150 °C		Al_2O_3 heated 2 h at 300 °C	
	B	S	B	S	B	S	B	S	B	S
1·50	13·4	57·3	13·4	55·7	13·4	59·2	13·4	57·2	15·3	56·2
4·00	9·6	37·2	9·6	39·2	9·6	38·8	9·6	39·8	13·4	45·5
5·50	3·8	18·0	3·8	17·5	3·8	17·2	3·8	17·3	3·8	21·1

Tab. 2. Dependence of ^{99m}Tc elution efficiency (%) on Al_2O_3 sterilisation procedure used
Volume of eluence: 5 ml 0·9% NaCl; Flow rate: 20 ml min^{-1} cm^{-2}; Column: lg Al_2O_3 (Woelm-R)

Al_2O_3 unsterilised	Al_2O_3 irradiated $2{\cdot}5 \cdot 10^4$ Gy heated 2h at 150 C°	Al_2O_3 irradiated $2{\cdot}5 \cdot 10^4$ Gy	Al_2O_3 heated 2 h at 150 °C	Al_2O_3 heated 22 h at 150 °C	Al_2O_3 autoclaved 0·5 h at 120 °C	Al_2O_3 autoclaved 2 h at 300 °C
93·7	84·7	84·8	90·6	89·0	89·8	91·3

autoclaving. By applying direct inoculation to corresponding nutrient substrates all the samples were found to be sterile.

By determining adsorption capacities of sterilised Al_2O_3 samples, and by comparing the obtained values with those for unsterilised samples, it has been found that the adsorption capacity of Al_2O_3 remains unchanged with respect to molybdenum, independent on the sterilisation procedure used (radiation or thermal).

By investigating ^{99m}Tc elution efficiency the highest value was obtained for unsterilised Al_2O_3 samples (94%), slightly lower for thermally sterilised and autoclaved (90%) and the lowest for gamma radiation sterilised Al_2O_3 (85%). The pH of Al_2O_3 suspension in H_2O remained unchanged, i.e., independent on the sterilisation procedure used.

Received August 8, 1984

References

[1] *H. Arino, H. H. Kramer,* Int. J. Appl. Radiat. Isotopes **26** (1975) 301

[2] *Lj. M. Jaćimović, S. M. Milenković, J. L. Vučina,* Isotopenpraxis **20** (1984) 69

[3] *J. Steigman,* Int. J. Appl. Radiat. Isotopes **33** (1982) 829

[4] United States Pharmacopeia XIX, 710

[5] *V. Jovanović, B. Zmbova, J. Bzenić,* Radiol. Jugosl. **13** (1979) 53

Isotopenpraxis **21** (1985) 6, S. 220 bis 223

Radiometric Measurements of Moisture and Density in Spruce Chips

U. Korell, H. Kurth, H.-J. Mette

(Dresden Technical University, Forestry Section)[1]

Not only an increased supply of rawwood and woood residues in the form of chips, but also aspects concerning the economic use of material in the enterprises of the woodworking indstry make questions relating to the techniques and methods of measurement an issue of particular interest for specialists on behalf of their high accuracy. In order to elucidate practicable methods for this purpose investigations were carried through using commercial immersion probes, which function on the principle of scattered radiation, and enclosed γ sources operating on the principle of transmission in chips of various dimensions and origin. On the one hand the task consisted in the elaboration of calibration curves and in the collection of factors, which exert an influence on the accuracy of measurement, on the other.

Ein erhöhtes Angebot an Rohholz und Holzresten in Hackschnitzelform, aber auch materialökonomische Gesichtspunkte in den Betrieben der holzverarbeienden Industrie begründen eine steigende Nachfrage nach Meßverfahren hoher Genauigkeit und Reproduzierbarkeit. Dazu wurden mit handelsüblichen Tiefensonden nach dem Prinzip der Streustrahlung und mit umschlossenen γ-Strahlern nach dem Prinzip der Durchstrahlung an Hackschnitzeln unterschiedlicher Größe und Herkunft Untersuchungen durchgeführt. Dabei ging es einmal um die Erarbeitung von Kalibrierkurven, zum anderen um eine möglichst numerische Erfassung von Faktoren, die die Genauigkeit eines Meßergebnisses beeinflussen.

Keywords

densimeters; density; measuring instruments; moisture gages; pines; radiometric gages; wood

1. Introduction

In forest exploitation and wood working factories the chipping method is applied in order to give the delivered material (small-sizes wood and wood residues), which in most cases is bulky, a suitable form for its manipulation. However, this technological advantage is partly set aside inasmuch as calculating the quantity of chips in cubic metres of rawwod (V) by means of the conversion factor (u) on the basis of the bulk volume (V_R) to be measured leads to inaccuracies, which are unacceptable from the stnapoint of economical use of the material

$$V = V_R \cdot u \,, \qquad m^3$$

u varies strongly depending upon tree species, moisture content and dimensions of the chips as well as chip type and loading technique so that it seems nearly impossible even when working with utmost cara to obtain generally adopted mean conversion factors. In addition to this one has also to reckon with uncertainties due to compaction of the chips during transport.

If increased quantities of rawwood and wood residues in chip form have to be used suppliers and purchasers face the problem of finding an accurate and fast method for dry matter determination. Radiometric methods should be given priority in this field as already used in civil engineering, in the meteorological service, in geophysics, lignite mining, water management and agriculture. At the Department of Forest Utilization (Dresden Technical University, Tharandt Forestry Section) first fundamental investigations were undertaken in 1974 in order to test the practical applicability of this method also in the case of chips.

2. Measurement with application of scattered radiation

For calculating the relative dry matter content of a certain amount of chips the following known relations are applied:

$$T = \frac{m_{dtr}}{m_f}$$

$$m_{dtr} = m_f - w$$

$$T = \left(1 - \frac{w}{m_f}\right)$$

[1]) Address: DDR-8223 Tharandt, Pienner Str. 8

In the radiometric measuring method, w can be substituted by w' and m_f by m_f' and the equation reds:

$$T = \left(1 - \frac{w'}{m_f'}\right)$$

Here, T proportion of dry matter; m_{dtr} dry mass of chips in kg; m_f green mass of chips in kg; w water mass in kg; m_f' green mass in kg/m³ measured with γ probe according to $m_f' = a + b \cdot I_\gamma$[2]); w' proportion of water in kg/m³ measured with n probe according to $w' = a' + b' \cdot I_n$[3]).
(see Fig. 1).

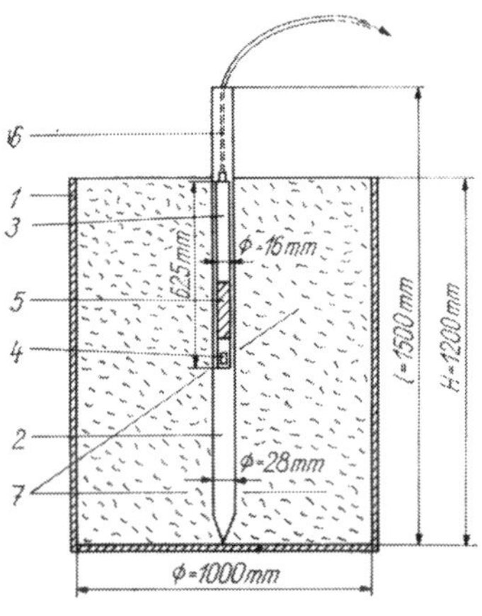

Fig. 1. Measuring moisture and density in chips by scattered radiation in a wooden barrel

1 – wooden wall 24 mm thick; 2 – immersion tube of iron (for holding the measuring probe); 3 – measuring probe; 4 – radiation source (n probe wiht Am-Be 2·96 GBQ; probe with ¹³⁷Cs 0·75 GBQ; 5 – detector (n probe with BF3 radiation detector NUM 712 A 20; probe with halogen counter CTC-5-STS-5); 6 – cable leading to radiation meter 20026; 7 – chips as material to be measured

For radiometric determination of the dry matter content the following regression equations, which were calibrated with the immersion probes S 23[4]) and S 24[5]) [1, 2], are adopted

$$w' = 0{\cdot}004505 \cdot I_n - 37$$

$$m_f' = 0{\cdot}004652 \cdot I_\gamma - 239$$

For an average content of water equalling $\overline{w}' = 120$ kg/m³ and for an average green density equalling $\overline{m}_f = 283$ kg/m³ the errors at a confidence level of 0·99 amount to ±8 kg/m³ (±6·7%) and 17 kg/m³ (±5·5%), respectively. In general, an influence of the chip dimensions, i.e. of the structural composition, could not be revealed. However, for the investigated fractions R 4[6]), R 10, R 20 and R 30 this phenomenon occurs unrestrictedly in measurements of moisture. In measurements of density a high proportion of fine fractions (R 4) seems to exert an unfavourable effect on the result of measurement. In general, however, the proportion of fine fractions is low and does scarcely play any role in error calculation.

[2]) I_γ γ-pulse density/min

[3]) I_n n-pulse density/min

[4]) Neutron immersion probe S 23 (neutron source Am-Be, type ANZ 80, activity about 3 × 10⁹ Bq, dose rate 10⁻³ mGy/h at a distance of 1 meter from the operating container)

[5]) γ immersion/probe S 24 (γ source ¹³⁷Cs, type LAA/TGL 17997, activity when delivered about 0·75 · 10⁹ Bq, dose rate about 0·001 mGy/h at a distance of 1 meter from the operating container)

[6]) Sieve residues after fractionation
4 mm < R 4 < 10 mm < R 10 < 20 mm < R 20 < 30 mm < R 30

3. Measurement with the aid of transmission

From the viewpoint of wood sicnece uncertainties in radiometric measurements can be expected, as far as one knows, as a result of different densities of the wood as an absorber. These uncertainties are due to sometimes heavy fluctuations depending, e.g. upon tree section, knottiness and age. For precise determination of such uncertainties measurements with transmission were carried through (higher measuring accuracy with by far less chip quantities compared with former investigations on the basis of scattered radiation – 6 kg against 200 kg). For this purpose 60 trees of a 31-year-old spurce stand, which had been remarked for thinning, were cut in the Tharandt forest district belonging to the Tharandt demonstration forest factory (near Dresden). From each of these 60 trees stem disks were taken from the central parts of the bottom, middle and top section and the pieces left were chopped.

In investigations dealing with the influence of wood density the basic density was taken as a basis $R = m_{dtr}/V_u$ in kg/m³ (m_{dtr} weight of oven-dry wood in/kg; V_u green volume of completely swollen wood in m³).

The results of the investigation were evaluated in compliance with the χ^2-test and in accordance with variance analysis.

3.1. *Measuring stem disks*

The basic density (see Tab. 1) in bottom sections in normal cases (418 kg/m³) significantly differs from those in the central and top sections (384 kg/m³ and 378 kg/m³ respectively). Depending upon knottiness there is a significant difference between the basic densities in stem disks with coarse and medium-sized knots as well as knot-free stem disks (420 kg per m³, 397 kg/m³, 383 kg/m³).

To ascertain the influence of tree section and knottiness on the result of raiometric measurement attenuation coefficients were determined on the basis of stem disks (see Tab. 2) on account of the low thickness of these stem disks with the aid of Am-241 as a low-energy radiator (see Fig. 2).

Tab. 1. Mean basic density "R" in kg/m³ depending upon stem section (F bottom; M central; Z top) and knottiness (A0 knotfree; A1 with medium-sized knots; A2 with coarse knots)

knottiness	stem section						$\overline{R}$	sa. N
	bottom		central		top			
	R	N	R	N	R	N		
A0	403	22	379	38	361	17	383	77
A1	442	6	379	7	370	8	397	21
A2	430	15	410	6	413	9	420	30
$\overline{R}$	418		384		378			
sa. N		43		51		34		128

According to the already mentioned statistical gest methods the linear attenuation coefficients μ show significant differences depending upon the abovementioned features ($\mu_F' - \mu_Z' = 0{\cdot}010$ cm⁻¹; $\mu_F' - \mu_M' = 0{\cdot}005$ cm⁻¹; $\mu_{A0}' - \mu_{A2}' = 0{\cdot}007$ cm⁻¹).

As expected, the small differences in the mass attenuation coefficients $\frac{\mu'}{R}$ are not significant.

Tab. 2. Mean linear attenuation coefficients μ' (cm^{-1}) and mass attenuation coefficients $k' = \frac{\mu'}{R}$ (cm^2/g) for stem disks depending upon

a) stem section	N	μ'	$s_{\mu'}$	k'	$s_{k'}$
bottom	43	0·077	±0·012	0·186	±0·027
central	51	0·072	±0·009	0·187	±0·021
top	34	0·067	±0·010	0·180	±0·023
b) knottiness					
A0	77	0·071	±0·010	0·184	±0·024
A1	21	0·073	±0·009	0·185	±0·018
A2	30	0·078	±0·013	0·185	±0·027

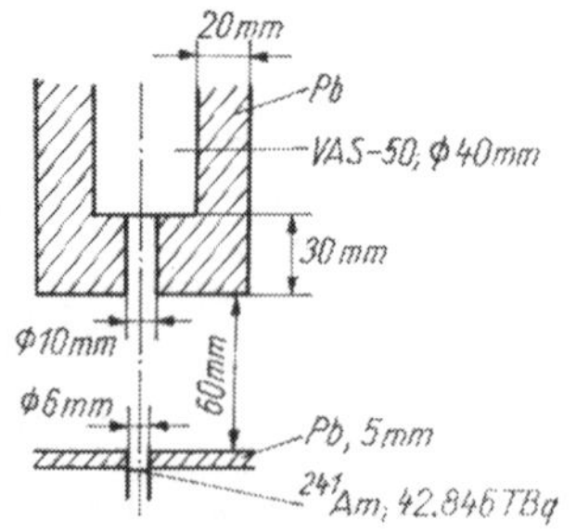

Fig. 2. Measuring density in stem disks by transmission

3.2. *Investigations carried out on chips*

The investigations on chips from different tree parts (bottom, central, top section) were carried out with the aid of ^{137}Cs (see Fig. 3). In the main they confirm the results obtained on the basis of stem disks (see Tab. 3) and exhibit a tendency towards a decrease of the attenuation coefficients μ from bottom to top corresponding to different basic densities. A significance of these differences could not be ascertained. The mass attenuation coefficient $\left(\frac{\mu}{R \cdot u}\right)$ is independent of the tree section and in its magnitude similar to that of water.

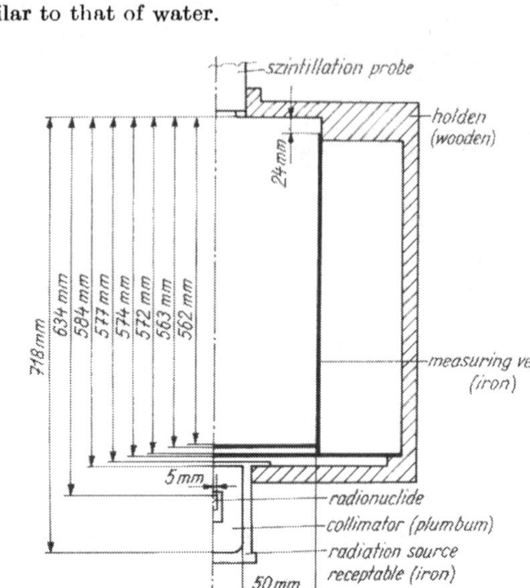

Fig. 3. Direct measurements of chips by transmission

Tab. 3. Attenuation coefficients for chips R 10 depending upon stem section

stem section	μ, cm^{-1}	$k = \frac{\mu}{R \cdot u}$, cm^2/g
bottem	0·0143	0·084
central	0·0140	0·084
top	0·0137	0·083

4. Discussion of results and conclusions

On the grounds of the basic density determination in stem disks it is possible to confirm values and relations known from literature [4, 5, 9]. This applies also to the mean basic density of 394 kg/m³, which is calculated on the basis of section-dependent densities ($R_F = 414$ kg/m³; $R_M = 384$ kg/m³; $R_Z = 378$ kg/m³) and of different portions of wood from the bottom, central and top section (50%; 40%; 10%). A change in the quantitative composition of the stem parts necessarily leads to deviations from this mean value. Knowledge of such changes becomes a topic of interest particularly in cases, in which quantity determinations in m³ of rawwood have to be effected on the basis of its green weight, because this must be practised in sale by weight as long as rawwood prices are related to weight.

The results of the investigations justify a subdivision of the wood into only two knottiness gradations (knotty, knotfree), especially as a subdivision into three knottines gradations is mathematically difficult.

As expected, depending upon $\varrho = R \cdot u$ (R basic density in kg/m³, $u = 0{\cdot}425$ bulk factor) the linear attenunation coefficient "μ", which was experimentally determined with ^{137}Cs in chips, is in good harmony with the corresponding series of measurements for some materials (see Fig. 4) and also well reflects differences in the results of radiometric measurements, which are due to the density of wood.

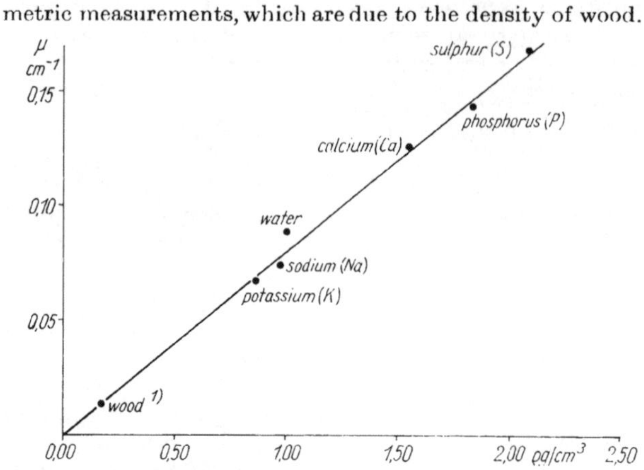

Fig. 4. Linear attenuation coefficients (μ) depending upon the density of some selected materials (ϱ)

[1]) $\varrho = R \cdot u$ (R basic density in g/cm³; u bulk factor)

As expected, depending upon the adopted nuclide the mass attenuation coefficients $\frac{\mu}{\varrho}$ are different, but equal within a series of measurements. Therefore, $\mu' = \frac{\mu'}{R}$ and $\mu = \frac{\mu}{R \cdot u}$ may serve also as important parameters in determinations of moisture and dry matter content inasmuch as with their aid heavy and variously caused fluctuations of density can possibly be levelled. The mass attenuation coefficient amounting to $k = 0{\cdot}084\ cm^2/g$, which was experimentally determined in chips with the aid of ^{137}Cs, is in good accordance with the theoretical value for conifer wood (0·086 cm^2/g).

The reciprocal values of the attenuation coefficients can be a good indicator of the layer thicknesses and areal weights respectively to be chosen with the aim of reaching a measuring sensitivity, which is as optimal as possible. For oven-dry wood and if adopting ^{241}Am they amount to 14 cm and 5 g/cm^2 respectively, when using ^{137}Cs the corresponding values are 70 cm and 12 g/cm^2.

Received February 29, 1984
Accepted in revised form May 16, 1984

References

[1] Bedienungsanleitung für die Neutronen-Tiefensonde S-23.1 Bauakademie der DDR, Institut für Industriebau, Dresden 1982
[2] Bedienungsanleitung für die Gamma-Tiefensonde S-24. 2. (1982) Bauakademie der DDR, Institut für Industriebau, Dresden 1982
[3] *K. Irmer, S. Richter, S. Koch,* Isotopenpraxis 2 (1966) 131
[4] *W. Knigge, H. Schulz,* Grundriß der Forstbenutzung. Hamburg/Berlin(West): Verlag Paul Parey 1966
[5] *F. Kollmann,* Technologie des Holzes und der Holzwerkstoffe. Bd. 1. Berlin (West), Göttingen, Heidelberg, München: Springer-Verlag 1951
[6] *U. Korell, O. Bloßfeld,* Die Anwendung radiometrischer Methoden für Dichte- und Feuchtemessung an Hackschnitzeln. In: Wissenschaftliche Tagung der Sektion Forstwirtschaft vom 16. bis 18. Oktober 1978. – TU Dresden, Sektion Forstwirtschaft – Band III. Dresden und Tharandt 1978, S. 109 – 116
[7] *H. Kurth,* Ein Beitrag zur radiometrischen Feuchte- und Dichtemessung an Fichtenhackschnitzeln und Sägewerksabfällen aus Fichte. Dresden: TU, Sektion Forstwirtschaft, Diplomarbeit 1982, unveröffentlicht
[8] *G. Langendorf, E. Schuster, R. Wagenführ,* Rohholz. Leipzig: Fachbuchverlag 1966
[9] *H.-J. Mette, U. Korell,* Richtzahlen und Tabellen für die Forstwirtschaft. Berlin: Deutscher Landwirtschaftsverlag 1972
[10] *E. Oftedal,* Norks Skogind, Oslo 32 (1978) 206
[11] *R. Storm,* Wahrscheinlichkeitsrechnung, mathematische Statistik und statistische Qualitätskontrolle. Leipzig: Fachbuchverlag 7 1979
[12] TGL 25 106 Prüfung von Holz. – 1979

Kurze Originalmitteilungen

Isotopenpraxis 21 (1985) 6, S. 223 bis 225

Characterization of Al_2O_3 Ceramics by NAA and XFS Methods*)

H. Rausch, S. Török, A. Simonits

(Central Research Institute for Physics of the Hungarian Academy of Sciences, Budapest)[1]

Keywords

aluminium oxides; autoradiography; ceramics; microelectronics; multi-element analysis; nondestructive analysis; X-ray fluorescence analysis

Introduction

The impurities and their distributions have been investigated in Al_2O_3 based ceramic plates by Nondestructive Activation Analysis (NAA), Autoradiography and X-Ray Fluorescence Spectroscopy (XFS). The properties of these ceramics used in the electronic device production may be influced both by the contributed element oxides constituting the basis of the glass phase and by some essential metal oxides and trace element which may take a prominent part in the formation of the metal/oxide binding.

At the same time the ceramic samples turned out to be a well-defined suitable multielement medium to control the k_0-standardization technique [1, 2]. Besides, multielement standardization has also been used for all cirtical elements to compare the results obtained by the two methods.

Experimental

The samples to be investigated were 500 micron thick plates with a 2 cm^2 surface size. Neutron irradiations were performed in the KFKI WWRS-M reactor at an average thermal neutron flux of $3{\cdot}2 \times 10^{13}\ n \cdot cm^{-2} - s^{-1}$. The flux ratios, φ_s/φ_e, were 18·3 and 31·6 in the irradiation channels used, measured by a 100 micron thick Zr foil. In order to determine a possible maximum number of components, largely different irradiation periods (60 s, 3600 s and $1{\cdot}44 \times 10^5$ s) were used. By short irradiation, when the irradiation capsule was previously cooled with liquid nitrogen, the flux ratio increased from 31·6 to 50·9. For the activity measurements a gamma spectrometer consisting of a Canberra Ge(Li) detector of 0·78 keV per 122·07 keV and 1·82 keV/1332·5 keV resolution (efficiency 13%), a Canberra 2010 type amplifier and associated electronics and a KFKI ICA-70 type 4K channel analyser was used.

The gamma spectra were evaluated semi-automatically by the program HYPERMET [3] adapted for an IBM-360 compatible computer. Quantitative evaluation is performed by a PDP 11/40 compatible computer with a program using a disc-based nuclear data library. In addition, multielement standards were also prepared for several elements grouped as (Ca, Ba, Fe, Co), (Ti, Ni, Zn, Ga) and (Nd, Lu, Yb, Eu). Preparation was performed from standard liquid solutions by pipetting and adsorbing of 20 to 50 µl aliquots in spectral-pure carbon vessels.

The measured spectra were disturbed by the presence of high energetic matrix activity produced of ^{28}Al, ^{27}Mg and ^{24}Na induced in the Al target. This effect could be reduced partly by repeating the measurements after different cooling periods and using different geometry. After cooling periods of about 10^5 s and 10^6 s autoradiographs were made on the surface of the activated plates using ORWO-Macro-Autoradiography film AF3-4. The density distributions of the autoradiograms were evaluated by a two-way microdensitometer.

*) Presented at the 3rd Symposium on "Nuclear Analytical Methods", April 11 – 15, 1983, Dresden, GDR

[1]) Address: H-1525 Budapest 114, P. O. Box 49, Hungary

The layout of the X-ray fluorescence spectrometer was the following: characteristic X-rays were excited by ^{55}Fe and ^{109}Cd annular radioisotope sources placed 5 mm from the detector. The X-rays emitted from the sample located 17 mm from the detector passed through the 25 micron Be window of a Canberra 7300 type Si(Li) detector of 30 mm² active area. The pulses of the preamplifier (model 1708) were amplified by a spectroscopy amplifier (model 1713). Pile-up was eliminated by a spectrum enhancer (model 1764). The analog signals were converted and stored in a Canberra 8100 multichannel analyser and evaluated by an off-line TPA-L microcomputer (KFKI). Simple Gaussian peaks and a least squares fit of polynomial background was applied to calculate net peak areas.

Discussion

The major steps of the applied k_0-standardization method can be summarized briefly as follows:

— The concentration of the element investigated may be expressed as:

$$c_i\,[\text{wt}\cdot\text{ppm}] = \frac{A_{p,i}\cdot F_{sdc}}{w\cdot k\cdot A^*_{sp}} \qquad (1)$$

where the asterisk refers to the comparator, w represent the sample wight in grams and $A_{p,i}$ the average count rate of the full-energy peak of the measured isotopes.

F_{sdc} is the correction factor for saturation, decay and measuring periods:

$$F_{sdc} = \frac{\lambda_i\cdot t_m}{(1-e^{-\lambda_i t_i})\,(e^{-\lambda_i t_d})\,(1-e^{-\lambda_i t_m})} \qquad (2)$$

According to the definition, $k = A_{sp}/A^*_{sp}$, represents here the relation of the specific count rate (per µg) of the isotope and comparator valid only for the given experimental set-up.

k can be calculated from Eq. (3):

$$k = k_{0,Au}\frac{f+Q_0}{f+Q_0^*}\cdot\frac{\varepsilon_p}{\varepsilon_p^*} \qquad (3)$$

where ε_p is the absolute full-energy peak efficiency of the detector. $Q_0 = I_0/\sigma_0$ and k_0 are nuclear constants determined experimentally with 1—2% accuracy by the Budapest-Gent Laboratories [4].

$f = \varphi_s/\varphi_e$ represents the flux ratio in the irradiation channel used and can be measured e.g. by co-irradiating a multiisotopic detector. In this work Zr foil was used. f can be calculated using the formula (4)

$$f = \frac{280\left[\dfrac{A_{sp}(^{95}\text{Zr, sum})}{A_{sp}(^{97}\text{Zr})}\right]\times 101{\cdot}4}{17{\cdot}58-\left[\dfrac{A_{sp}(^{95}\text{Zr, sum}}{A_{sp}{}^{97}(\text{Zr})}\right]} \qquad (4)$$

In many cases the epithermal neutron flux distribution may deviate from the ideal $1/E$ shape which I_0 and thus Q_0 are referring to. When representing the actual distribution with $1/E^{1+\alpha}$ function, corrections have to be applied in the case of $\alpha \neq 0$. The values of α may vary from $-0{\cdot}02$ to $0{\cdot}11$ depending mainly on the reactor moderator material and position of the irradiation channel.

Corrections for $Q_0(\alpha)$ values were made by means of the following Eq.

$$Q(\alpha) = \frac{Q_0(\alpha=0)-0{\cdot}429}{(\bar{E}_r)^\alpha}+\frac{0{\cdot}429}{(2\alpha+1)\,E^\alpha_{Cd}} \qquad (5)$$

Tab. 1.

Element	Concentration, wt · ppm					
	NAA[k_0-st]	σ, %	NAA[MES]	σ, %	XFS	σ, %
Ca	1934·0	3·5	2026·0	3·9	3800·0	32
Fe	453·3	2·1	441·8	2·2	610·0	14
Ti	—		296·4	11·0	320·0	13
Ba	24·6	3·5	23·8	3·3		
Zr	47·5	2·9	47·8	3·1	56·2	11
Mn	14·4	2·7				
Cu					12·0	27
Zn	6·78	1·8	6·64	2·2	4·2	24
Ga	12·5	3·7	13·0	4·1	17·0	16
Cr	5·34	3·9				
Nd	9·62	4·4	9·56	4·7		
Ce	0·714	3·0				
La	0·203	2·6				
Co	0·672	1·6	0·667	1·2	<1	
Ag	0·570	2·6				
U	0·219	4·2				
Th	0·196	4·1				
Hf	0·174	3·1				
Sc	0·044	3·2			<1	
Ni			0·857	4·4	2·1	24
Yb			0·017	10·0		
Lu			0·005	15·0		
Eu			0·018	10·0		
Ta	0·038	4·7				
Au	0·0042	2·4	0·0041	2·1		
Sr					15·0	6·6
Pb					3·2	12·5

The relevant α-values were determined by the triple monitor method [2] using, in addition to the Zr foil, an Al-alloyed gold wire. All corrections mentioned above were included in the program SPECTRUM [5] run on a PDP 11/4 compatible machine.

In quantitative X-ray fluorescence analysis the measured intensity $I_{F,i}$ is related to the elemental concentration c_i by the simple equation:

$$I_{F,i} = I_0 \cdot G \cdot \varepsilon(E_i) \cdot K_i \cdot C_i \cdot t(E_i) \quad (6)$$

where I_0 is the exciting X-ray intensity, G the factor of geometry, $\varepsilon(E_i)$ the detection efficiency for the characteristic radiation, K_i the fluorescence probability of the particular line. When trace analysis is the aim in a homogeneous layer of a low atomic number matrix, then secondary and tertiary fluorescence effects can be neglected and the absorption correction factor turns out to be very simple:

$$t(E_i) = \frac{1 - e^{-[\mu(E_i) + \mu(E_{ex})] \cdot m}}{[\mu(E_i) + \mu(E_{ex})] \cdot m} \quad (7)$$

where $\mu(E_i)$ and $\mu(E_{ex})$ are mass absorption coefficients of the entire sample for the characteristic and exciting radiation, respectively, and m is the thickness of the sample in mass per unit area. The sensitivity factors $[I_0 G \varepsilon(E_i) K_i]$ were measured by using infinitely thin targets consisting of high purity metals vacuum deposited on Mylar foils. For calculation of absorption corrections coefficients of *McMaster* et al. [6] are used and were checked by transmission experiment as described by *Giaque* et al. [7].

Results

Results obtained by the different methods are summarized in Tab. 1. Several observations can be made to the results compiled in the table. First, all data obtained by the relative method are in reasonable agreement with those calculated by the k_0-standardization technique. Due to the measured complex gamma-spectra (more than 50 peaks per spectrum) some peak areas could only determined with low precision thereby preventing any meaningful comparison. For some relevant elements Fig. 1 shows the deviation of the $c[k_0 - \text{st}]$ data from the $c[\text{MES}]$ ones accepted as reference. It can be seen that the deviations are within $\pm 5\%$.

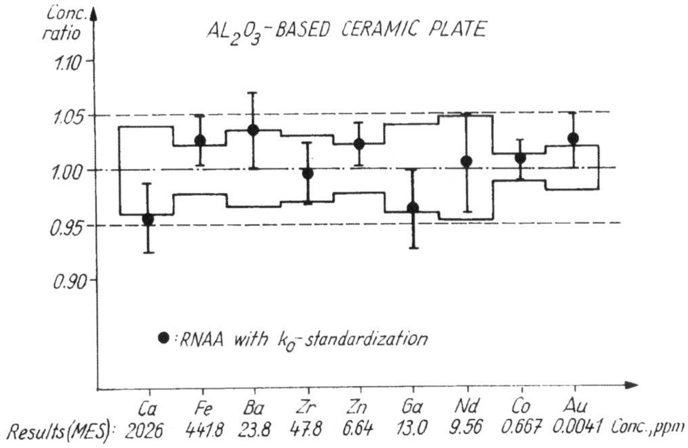

Fig. 1. Ratio of results from k_0-standardization to results obtained by relative standardization

For determination of Ti and Ni the fast neutron reactions of ^{50}Ti(n, α)^{47}Sc and ^{58}Ni(n, p)^{58}Co could be only utilized because of the background activity of the short-lived isotopes, Since ^{47}Sc are also formed during beta-decay of ^{47}Ca produced by the ^{46}Ca(n, γ) reaction, correction has to be applied calculated from the Ca content measured. Correction is negligible if the Ca content is about the same or less than the Ti concentration. No correction has to be made by Cd-covered irradiation.

Received April 18, 1984

References

[1] *A. Simonits, L. Moens, F. de Corte, A. de Wispelaere, A. Elek, J. Hoste*, J. Radioanal. Chem. **60** (1980) 461
[2] *F. de Corte, K. Sordo-El Hammami, L. Moens, A. Simonits, A. de Wispelaere, J. Hoste*, J. Radioanal. Chem. **62** (1981) 209
[3] *G. W. Phillips, K. W. Marlow*, Program HYPERMET for automatic analysis of gamma-ray spectra, NRL Memorandum Report 3198 (1976)
[4] *A. Simonits, F. de Corte, L. Moens, J. Hoste*, J. Radioanal. Chem. **72** (1982) 209
[5] *A. Demeter*, Priv. commun.
[6] *W. H. McMaster, M. Delgrande, J. H. Mallett, J. Hubbell*, University of California, Lawrence Radiation Laboratory Report, UCPL-50174 (1969)
[7] *R. D. Giaque, F. S. Goulding, J. M. Jacklevic, R. H. Pehl*, Anal. Chem. **45** (1973) 671

Isotopenpraxis 21 (1985) 6, S. 225 bis 228

Buchbesprechungen

**Radiological Assessment: Predicting the Transport, Bioaccumulation, and Uptake by Man of Radionuclides Released to the Environment
Bethesda, Maryland: Recommendations of the National Council on Radiation Protection and Measurements 1984**

300 S., 39 Abb., 61 Tab. (NCRP Report No. 76); kart.

Dieser Bericht mit empfehlendem Charakter gliedert sich in 7 Abschnitte, wobei die ersten 3 der Freisetzung von Radionukliden in die Atmosphäre, in das Süßwasser und in das Grundwasser gewidmet sind. Der 4. Abschnitt behandelt Folgen von radioaktiven Belastungen in den Nahrungsmitteln des Menschen, und die beiden folgenden Abschnitte stellen eine kritische Analyse von Modellrechnungen und Modellbetrachtungen dar, die gewisse Unsicherheiten zur Folge haben und bei notwendigen Schlußfolgerungen unbedingt zu beachten sind. Die formulierten Empfehlungen der NCRP, ein Glossar über auftretende Begriffe und das umfangreiche Literaturverzeichnis schließen diese Schrift ab.

Die hohe Nachweisempfindlichkeit von Radionukliden in der Umwelt des Menschen hat bereits eine lange Tradition in der modellmäßigen Erfassung und Abschätzung ihrer Folgeerscheinungen auf den Menschen. Es ist hervorzuheben, daß in dieser Schrift eine sehr sorgfältige Zusammenstellung solcher Modelle — ihrer Vorteile und Nachteile — vorliegt, die auch für nicht-radioökologische Überlegungen sehr wertvoll ist. Damit ist diese Schrift auch über das eigentliche Anliegen hinaus allen Wissenschaftlern zu empfehlen, die sich mit Problemen der Umweltchemie befassen.

H. Hübner

Radioaktive Stoffe und ionisierende Strahlung Quellen, Wirkung, Wertung Fachtagung des Deutschen Atomforums, 27.–28. Sept. 1983, Bonn

Hrsg.: Deutsches Atomforum e. V., Bonn 1984, 354 S.; karton.

Es handelt sich bei dem vorliegenden Buch um die Zusammenfassung von Vorträgen einer Fachtagung des Deutschen Atomforums. Wie im Untertitel hervorgehoben, geht es um „Analysen, Informationen und Stellungnahmen für die friedliche Nutzung der Kernenergie".

Die Thematik der Tagung war in drei Komplexe geteilt:

Strahlenexposition: Eine Bilanz
Strahlenrisiko: Quantifizierung und Wertung
Strahlenschutz: Formalismus und Umwelt.

Im ersten Komplex wurde neben einer zusammenfassenden Bilanz der Strahlenexposition die Strahlenbelastung aus natürlichen und künstlichen Strahlungsquellen, durch die Medizin und durch kerntechnische Anlagen behandelt.

Der zweite Komplex enthält Beiträge zum Verständnis von Dosis und Risiko, zu Strahlenbelastung und Risikobelastungen, zum technischen Fortschritt und Risiko und zum Vergleich der verschiedenen Lebensrisiken.

Im dritten Komplex wird untersucht, ob es einen Schwellenwert für Karzinogene gibt; weitere Ausführungen betreffen die Strahlenbelastung bei schweren Unfällen, Kosten und Nutzen im technischen Strahlenschutz sowie Fragen der Radioökologie. Ein Beitrag befaßt sich mit der Stillegung von Kernkraftwerken als einem gelösten Strahlenschutzproblem.

Das Buch kann allen Kollegen empfohlen werden, die mit Problemen des Strahlenschutzes zu tun haben.

K. Hübner

Use of nuclear techniques in the mineral industry (IAEA regional training course)

J. S. Watt, B. D. Sowerby (Eds.) Austral. Atomic Energy Commission, Research Establ., Lucas Heights, 1982 (AAEC/S24). VII + 471 S., 209 Abb., 43 Tab., 145 Lit.

Im Jahre 1980 führte die Australian Atomic Energy Commission (AAEC) gemeinsam mit der Commonwealth Scientific and Industrial Research Organization (CSIRO) einen Lehrgang über die Anwendung von kernphysikalischen Meßmethoden in der Grundstoffindustrie durch, der von der IAEA für die asiatische und pazifische Region vorgesehen war. Der Lehrgang fand in Australien statt und wurde von der australischen Regierung finanziell unterstützt. Der Lehrgang umfaßte 2 1/2 Wochen Vorträge und experimentelle Übungen in Lucas Heights, einen dreitägigen Kursus über die Prozeßkontrolle in der Rohstoffanreicherung an der Universität zu Brisbane und schließlich 2 Wochen Betriebsbesichtigungen von kernphysikalischen Meßsystemen in Bergbau- und Aufbereitungsbetrieben.

Die vorliegende Veröffentlichung enthält die Texte der Vorträge, die den Lehrgangsteilnehmern geboten wurden. Entsprechend dem Ziel des Lehrganges, der sich an Teilnehmer mit physikalischer oder technischer Hochschulbildung wandte, ist der Umfang der Vorträge vorwiegend auf die meßtechnischen Grundlagen und die praktischen Anwendungen gerichtet. Die einzelnen Kapitel behandeln: kernphysikalische Grundlagen (50 S.), Nachweis und Messung von Kernstrahlung (72 S), Gamma- und Neutronen-Quellen (14 S.), radiometrische Kontrolle (12 S.), Röntgen-Fluoreszenz-Analyse (78 S.), Analyse und Probennahme von rohen Schüttgütern (82 S.), Bohrlochmessungen (92 S.), Anwendungen von radioaktiven Tracern (38 S.), Strahlenschutz (33 S.). Hieraus ist bereits zu erkennen, daß alle praktisch interessierenden Meßaufgaben behandelt werden. Als Autoren werden 16 Wissenschaftler genannt, die in ihren Spezialgebieten schon länger durch Veröffentlichungen bekannt sind. Zu jedem Abschnitt werden anschauliche Abbildungen, weiterführende Literaturhinweise und aussagekräftige Tabellen geboten; man spürt, daß erfahrene Fachleute am Werk sind. Hervorzuheben ist der gut lesbare Schriftgrad des Bandes.

Wer sich in das genannte Gebiet einarbeiten will oder einen ähnlichen Lehrgang vorzubereiten hat, wird in diesem Werk eine ausgezeichnete Anleitung finden.

K. H. Köhler

Protection in Nuclear Medicine and Ultrasound Diagnostic Procedures in Children

National Council on Radiation Protection and Measurements, Bethesda, USA 1983. 81 S., 15 Tab., 1 Anhang, 90 Lit. (NCRP Report No. 73)

Der Titel der amerikanischen NCRP-Empfehlungen weist auf einen Entwicklungsstand hin, der in unserem Lande noch erreicht werden muß. Gegenwärtig gibt es ernsthafte Bemühungen, Untersuchungsverfahren, bei denen ionisierende Strahlung angewendet wird, durch Ultraschall zu ersetzen. Das ist besonders für Kinder und Jugendliche von Bedeutung. Strahlenphysiker, Radiologen und Pädiater werden von diesem Report direkt angesprochen.

In textlich äußerst knapper Form wird zunächst auf die Grundzüge der Untersuchungstechniken einschließlich der Geräteanforderungen in der Nuklearmedizin eingegangen (Sedierung, Injektionstechniken, Parameter von Scannern und Gamma-Kameras). Die Nutzen-Risiko-Betrachtungen sind allerdings ohne Substanz, trotz der ICRP-Reports Nr. 26 und 27, die nicht einmal erwähnt wurden. Der Leser muß über moderne inkorporationsdosimetrische Methoden informiert sein, um die Ausführungen zu verstehen. Wesentlich ist aber der Anhang. Diese 29 Seiten sind für den Praktiker von großem Wert. In Altersabhängigkeit vom Neugeborenen bis zum Alter von 5 Jahren werden die wachstumsbedingten anatomischen Veränderungen in den Angaben für die mittlere Dosis bezogen auf die kumulierte Aktivitätseinheit (keine SI-Angaben!) berücksichtigt. Das geschieht für die heute wesentlichen Radionuklide ^{51}Cr, ^{67}Ga, ^{99m}Tc, ^{111}In, ^{123}J, ^{131}J und ^{201}Tl. Von 6 bzw. 9 aktivitätsreichen Organen wird die Dosisberechnung für 24 Zielorgane ermöglicht. Im zweiten Teil des Berichtes werden zunächst mögliche Schädigungen durch Ultraschall beschrieben, die aber praktisch für diagnostische Anwendungen nicht relevant sind. Für einige, besonders in den USA hergestellte Ultraschallgeräte, werden Schalleistungsparameter angegeben, allerdings fehlen wesentliche Gerätefamilien, die den Weltmarkt mitbestimmen (z.B. Japan, BRD).

Ausführlich werden Literaturstellen zitiert, die retrospektive Studien enthalten und den Schluß erhärten, daß bisher im Bereich der in der Ultraschalldiagnostik angewandten Frequenzen und Schalleistungen pro Flächeneinheit keine Gewebeschäden am Patienten und am Untersucher gefunden wurden.

Insgesamt ein Report, der nicht vom Text sondern von den Tabellen her seinen Wert besitzt. Am Schluß kann sich der Leser über die Arbeitsweise und die Aktivitäten der NCRP informieren und erhält einen Überblick über die bisher erschienenen NCRP-Berichte. *M. Tautz*

Isotope and Radiation Techniques in Soil Physics and Irrigation Studies 1983
Proceedings of a Symposium Aix en Provence 18–22 April 1983, org. I.A.E.A./F.A.O.

Vienna: IAEA 1983. 596 S., zahlr. Abb., Tab. und Lit.-Angaben; kart. Austr. Sh. 1160.–

Eine für hydrologische, hydrogeologische, wasserwirtschaftliche, land- und forstwirtschaftliche Aufgabenstellungen sehr wesentliche Frage ist die der Wasserbewegung und -speicherung und des Stofftransportes im Boden. Diesem Fragenkreis galt ein Symposium, dessen Beiträge in diesem umfangreichen Sammelband nun vorliegen. Dabei geht es um den Einsatz von Isotopen- und Strahlentechnik zur Klärung der Vorgänge um das Bodenwasser, um Bodenphysik und insbesondere Bewässerungsstudien. Die besondere Aufmerksamkeit galt deshalb den Möglichkeiten der Nutzung von Isotopen und von Strahlentechniken zur Klärung und Bestimmung von Bodenwassergehalt, Hydrodynamik des Bodenwassers, Bodenwasserqualität und salzbelasteten Böden, auch der Migration von Pestiziden, Schwermetallen und Transuranen, Stickstoff u. a. Schließlich werden Methoden vorgestellt und diskutiert; insgesamt liegen 43 Beiträge vor.

Grundtenor der meisten Beiträge ist der Einsatz von Neutronen-Sonden zur Bodenfeuchtebestimmung in ungesättigten Böden und deren Interpretation auf der Grundlage des *Darcy*schen Gesetzes, also der hydraulischen Leitfähigkeit in Kombination mit der Stofftransportgleichung. Hier werden zahlreiche praktische und theoretische Untersuchungen vorgestellt, die in Lysimeterversuchen oder Felduntersuchungen an unterschiedlichen Böden und in unterschiedlichen Klimabereichen (Untersuchungen in Entwicklungsländern) durchgeführt wurden. Allerdings ist auch ein hoher Prozentsatz von Beiträgen in diesen Symposiumsband aufgenommen worden, die keinen unmittelbaren Bezug zur Isotopen- oder Strahlentechnik haben, die aber die Gesamtproblematik zur bodenphysikalischen Charakteristik oder Kennwertermittlung sowie zu Bewässerungsproblemen (hier insbesondere in Entwicklungsländern) abrunden. Wesentlicher Gegenstand ist immer wieder die Versalzung der Böden durch Irrigation und deren Begegnung.

Methodisch befassen sich die Beiträge vorwiegend mit der Neutronenaktivierung (Americium-, Plutonium-Quellen), weniger mit ^{60}Co, ^{15}N, D, ^{18}O, T. *H. Jordan*

Isotopes in Organic Chemistry
Vol. 6: Isotopic Effects, Recent Developments in Theory and Experiment

E. Buncel, C. C. Lee (Eds.). Amsterdam – Oxford – New York – Tokyo: Elsevier 1984, 265 S., 47 Abb., 27 Tab.; geb. US $ 88.50; Dfl 230.00

Die Untersuchung von Isotopieeffekten ist seit den 60er Jahren ein unentbehrliches Hilfsmittel für den Organiker. Die Diskussionen und Interpretationen von Reaktionsmechanismen können subtiler und detaillierter geführt werden, bilden doch die Vergleiche von berechneten und experimentell ermittelten Isotopieeffekten den einzigen Zugang zur Aufklärung der Struktur von Übergangskomplexen. In zunehmendem Maße werden diese Interpretationen auf die anderen Zweige der Chemie bis hin zur technischen Chemie ausgedehnt und bilden auch das notwendige Rüstzeug für Studierende der Chemie.

Nach einer Zeitspanne von 4 Jahren erschien in der Reihe „Isotope in der organischen Chemie" von *E. Buncel* und *C. C. Lee* der vorliegende 6. Band. Er beschäftigt sich mit einigen weniger bekannten Isotopieeffekten, die aber ebenso wie die Isotopieeffekte, die nur auf den Massenunterschieden der Isotope eines Elements beruhen, einen hohen Grad an Beispielhaftigkeit und Übertragbarkeit haben. Da werden zunächst die Isotopieeffekte in den ^{13}C-NMR Verschiebungen und Kopplungskonstanten von *D. A. Forsyth* (Boston) behandelt, wie sie durch D, ^{18}O, ^{15}N und andere Nuklide verursacht werden können. Beispielsweise können durch das ^{18}O-Nuklid ^{13}C-Verschiebungen beobachtet werden, die zur Verfolgung einer Markierung geeignet sind, obwohl ^{18}O selbst NMR-inaktiv ist. In einem 2. Beitrag von *N. S. Isaacs* (Reading) werden die vieldiskutierten Einflüsse des Druckes auf die kinetischen Isotopieeffekte dargestellt. Da die Schwingungsfrequenzen gegenüber Druckänderungen verhältnismäßig unempfindlich sind (die O-D-Streckschwingung von 2100 cm^{-1} in wäßriger Lösung nimmt nur um ungefähr 2,3 cm^{-1} × kbar^{-1} ab), bleiben nur die präexponentiellen Faktoren übrig, um solche Effekte zu interpretieren. In einem 3. Beitrag werden die Magnetischen Isotopieeffekte von *N. J. Turro* (New York) und *B. Kräutler* (Zürich) übersichtlich dargestellt; hohe magnetische Feldstärken können bei Radikalreaktionen zu recht beachtlichen Isotopieeffekten führen, die allerdings außerhalb der bekannten Theorie der Isotopieeffekte als Masseneffekte betrachtet werden müssen. Schließlich wird in einem letzten Abschnitt von *L. B. Sims* (Raleigh, North Carolina) und *D. E. Lewis* (Waco, Texas) die Methode der Berechnung von Isotopieeffekten nach der Bindungsordnung ausführlich beschrieben; ein Weg, der häufig mangels anderer Daten gegangen werden muß und i. allg. recht vielversprechend ist.

Hier liegt eine moderne Zusammenstellung und Auffassung über Isotopieeffekte in der organischen Chemie vor, die eine Lücke füllt und neue Möglichkeiten und Wege aufzeigt. Daß man sich noch weitere Kapitel vorstellen könnte – Methoden der quantenchemischen Störungsrechnung oder quantenchemische Methode der *ab-initio*-Berechnungen von isotopomeren Molekülen –, kann nur am Rande vermerkt werden. Die Literatur ist mit 538 Zitaten reichlich verarbeitet worden, wobei allein aus den Jahren 1976–1984 41,5% stammen; eine hohe Aktualität, die dem Leser leicht weiterhelfen wird. Eine gute und sehr übersichtliche Darstellung der Anwendungsmöglichkeiten von Isotopieeffekten wird präsentiert, die eine weitreichende Benutzung und willkommene Aufnahme in den Kreisen der Organiker finden wird. *H. Hübner*

Laser Theory

H. Haken. Berlin – Heidelberg – New York – Tokyo: Springer 1984. XV + 320 S., 72 Abb.; kart. DM 90.–; approx. US $ 35.30

The "Laser Theory" of *Hermann Haken* is today considered a classic book by many laser physicists. It is now being republished in paperback. The fundamental aspects of the nature of laser light and many details of laser action at present seem to be well understood. Therefore the author has confined himself to a treatment of questions which are basic from the physical point of view. The book starts with a treatment of the theory of laser resonators, the quantum theory of coherence, and the quantization of electromagnetic fields.

The laser process can be fully understood only if the coupling of the light field and the laser atom to the surroundings is taken into account. Methods for dealing with this problem on the basis of extended field equations in quantum mechanically consistent form are presented in detail. Especially the quantum-mechanical treatment of laser light by means of quantum-mechanical *Langevin* equations, the density matrix equation, and the *Fokker-Planck* equation is presented.

After averaging over the fluctuations caused by the pumping and losses, the "semiclassical equations" are obtained, which can well be applied to the calculation of frequency shifts, populations, the photo-echo, ultrashort pulses etc.

Finally when all phase relations between lightfield and atoms are neglected, the author ends up with the rate equations, which have been playing an important role in the understanding of gross features of the laser, such as the threshold condition, output power, giant pulse laser, coexistence of modes and similar problems.

The corresponding chapters are written in such a way that they can be read independently from each other. In the last chapter several modern methods for dealing with quantum systems or from thermal equilibrium are discussed, for instance the general form of the density matrix equation which is not confined to the *Markoff*ian assumption and how one can establish an exact generalized *Fokker-Planck* equation for a c-number distribution function, which belongs to the general density matrix equation.

The volume is a valuable compendium, a source of first-rate information and a stimulus for those working in the field of laser theory as well as for those working in the field of laser photochemical excitation of molecules with respect to isotope separation. *H. Johansen*

Mikroelektronik für Einsteiger

R. Böker, G. Franke, A. Gärtner, A. Hunger, A. Kohl. 2. Auflage. Düsseldorf: VDI-Verlag GmbH 1984. 179 S., 104 Abb.; karton. DM 29,–

Nach Erscheinen der 1. Auflage 1983 wurde auf Grund des großen Erfolges bereits 1984 eine 2. Auflage erforderlich. Dabei wurden einige Druckfehler der 1. Auflage behoben. Ziel des Buches ist es, das interessante und hochaktuelle Themengebiet der Mikroelektronik dem technisch interessierten Leser nahezubringen, ohne daß dieser hierzu besondere Vorkenntnisse benötigt. Der Begriff Mikroelektronik bezieht sich dabei ausschließlich auf die Errungenschaften hinsichtlich der Mikrorechentechnik. Dieses Gebiet wird jedoch sehr übersichtlich in einer großen Breite behandelt.

Das Buch ist in 10 Kapitel gegliedert. Nach einer Einführung zur Problematik der Entwicklung der Mikroelektronik werden im 2. Kapitel Grundlagen der Signal- und Zahlenverarbeitung sowie zum allgemeinen Aufbau eines Mikrocomputers vermittelt. Die Kapitel 3, 4 und 5 beschäftigen sich mit den drei Hauptkomponenten Prozessor, Speicher, Peripherie eines Mikrocomputers. Dabei werden allgemeine Prinzipien erläutert und auch bestimmte mikroelektronische Schaltkreise als Beispiel erwähnt, ohne jedoch näher auf deren Beschreibung einzugehen, was den Rahmen dieses Grundlagenbuches übersteigen würde. Das Kapitel 6 gibt einen Einblick in Rechnerarchitekturen, wobei das Ziel das Verständnis der Verknüpfung der in den vorangegangenen Abschnitten erläuterten Komponenten ist. Hier wird auch auf den heute fließenden Übergang vom Mikrocomputer bis zu den Großrechnern hingewiesen. Im 7. Kapitel wird die Problematik der Programmierung gestreift, wobei speziell zu Programmiersprachen und Betriebsprogrammen grundlegende Aussagen getroffen werden. Das Buch schließt ab mit den Kapiteln Zusammenfassung und Ausblick, Bildnachweis und Sachwortverzeichnis. Letzteres erweist sich als sehr nützlich, um sich bezüglich bestimmter Fachvokabeln schnell einen Überblick zu verschaffen. Das Buch ist deshalb nicht nur dem interessierten Laien zu empfehlen, sondern auch allen Fachleuten, die beginnen, sich mit der Mikrorechentechnik zu befassen. *M. Steinert*

Integrierte Mikroschaltungen

Redaktion: B. V. Tarabrin. Moskau: Radio i Svjaz 1984. 528 S., vorwiegend Abb. und Tab. (in Russisch); geb. Rubel 2,–; M 10,–

Die Vielzahl der heute existierenden Halbleiterbauelemente, insbesondere der integrierten Schaltkreise, hat zu einem großen Bedarf an entsprechendem Datenmaterial geführt. Dabei steht der Anwender oft vor dem Problem, sich über Funktion, Anschlußbilder und Parameter von bestimmten integrierten Schaltkreisen zu informieren.

Das vorliegende Datenbuch sowjetischer Schaltkreise umfaßt sowohl analoge als auch digitale Typen bis hin zu Mikroprozessorfamilien. Leider werden keine Aussagen hinsichtlich der Kompatibilität zu vergleichbaren Schaltkreisen anderer Hersteller getroffen.

Das Buch ist in 4 Abschnitte gegliedert. Der erste Abschnitt ist allgemein gehalten, legt die Terminologie fest, liefert einen Typenschlüssel, gibt Gehäusebeschreibungen und dergleichen. Den digitalen Schaltkreisen ist der zweite Abschnitt gewidmet. Er bildet den Hauptteil des Buches. Von 47 Serien (!) werden die einzelnen Bausteine hinsichtlich logischem Schaltbild (nach RGW-Standard), Anschlußbelegung und elektrischer Parameter beschrieben. Beim Vergleich mit anderen Datenbüchern zeigte sich, daß auch das vorliegende Buch nicht frei von Druckfehlern ist. Das kann man aber bei einer solchen Vielzahl abstrakter Daten auch nicht erwarten. Der dritte Abschnitt enthält die Beschreibung der Bauelemente von 35 analogen Serien. Die Einbauvorschriften sind im 4. Abschnitt zusammengefaßt. Als sehr nützlich erweist sich der Anhang. Dort ist zum einen die Einteilung der Schaltkreise nach der Funktion vorgenommen und zum anderen wird hinsichtlich der verwendeten Logik klassifiziert.

Das Buch stellt für alle Ingenieure und technischen Kräfte, die mit sowjetischen integrierten Schaltkreisen arbeiten, eine wertvolle Hilfe dar, wobei es durchaus vorkommen kann, daß man den einen oder anderen Typ in diesem Datenbuch nicht findet. Auf Grund des Übersichtscharakters und der großen Breite wurde bei den höher integrierten Schaltkreisen auf detaillierte Angaben, wie sie der Entwicklungsingenieur benötigt, verzichtet. *M. Steinert*

Systems Analysis – Modelling – Simulation

Journal of Mathematical Modelling and Simulation in Systems Analysis

Edited at the Central institute of Cybernetics and Information Processes of the Academy of Sciences of the G.D.R. in cooperation with the Mathematical Society of the G.D.R. by Achim Sydow (Berlin) with the assistance of an international team of scientists

Published yearly, 1 volume (4 numbers).
Volume 2, 1985
16,7 cm by 24,0 cm, Annual list of contents.
Annual subscription M 120,—; single number M 30,—.
Order number: 1097 ISSN 0232-9298

The journal is devoted to art and techniques and applications of modelling and simulation in systems analysis. Systems analysis is an interdisciplinary activity. Articles will be published on systems analysis and systems theory, modelling and mathematical models and tools for decision and control, computer simulation systems, modelling and simulation in environmental protection, macroeconomics, agricultural production, regional planning, biosciences, traffic control, large engineering systems, resources distribution, management systems etc. In addition to original papers, survey articles and reports on conferences etc. are published. Finally, books covering the above mentioned and related subjects are reviewed.
Of interest to: Scientists of systems sciences, informatics, control engineering, applied mathematics, operation research, biology, economy, physics, environmental protection agriculture etc.

Orders can be sent

- in the FRG and Berlin (West): to the wholesale distributing agency Kunst und Wissen, Erich Bieber OHG, Wilhelmstr. 4—6, D-7000 Stuttgart 1;
- in the other Western European countries: to Kunst und Wissen, Erich Bieber GmbH, Dufourstr. 51, CH-8008 Zürich;
- in other countries: to the Akademie-Verlag, DDR-1086 Berlin, Leipziger Str. 3—4, PF-Nr. 1233.

AKADEMIE-VERLAG · BERLIN
DDR-1086-Berlin, Leipziger Straße 3—4

ISOTOPENPRAXIS

Band 21 | 1985 | Heft 6

Contents | Inhalt

►Language of Publication | ►Publikationssprache

Isotopenpraxis is indexed in Current Contents/PC & ES, Science Citation Index, and INIS Atomindex

FSC
www.fsc.org
MIX
Papier | Fördert gute Waldnutzung
FSC® C083411

Zeitfracht Medien GmbH
Ferdinand-Jühlke-Straße 7
99095 Erfurt, Deutschland
produktsicherheit@kolibri360.de